# HOW TO BE A HAM

# 3RD EDITION

# HOW TO BE A HAM

# 3RD EDITION

W. EDMUND HOOD
W2FEZ

TAB BOOKS Inc.
Blue Ridge Summit, PA

THIRD EDITION
SECOND PRINTING

Printed in the United States of America



Library of Congress Cataloging in Publication Data

Hood, W. Edmund.
How to be a ham.

Rev. ed. of: How to be a ham / by Ken Sessions and W. Edmund Hood. 2nd ed. 1981.
Includes index.
1. Radio—Amateurs' manuals. 2. Radio operators—Licenses—United States. I. Sessions, Ken E. How to be a ham. II. Title.
TK9956.H4574 1986 621.3841'66 85-27737
ISBN 0-8306-0353-0
ISBN 0-8306-0653-X (pbk.)



Questions regarding the content of this book should be addressed to:

Reader Inquiry Branch
TAB BOOKS Inc.
Blue Ridge Summit, PA 17294-0214

# Contents

# Preface

AMATEUR RADIO IS FOR EVERYONE. THERE ARE NO CLASS boundaries in ham radio; political, financial, and racial barriers crumble via the dial of the ham's rig. As a ham, you might find yourself carrying on a lively conversation with Barry Goldwater one day and/or some other famous celebrity the next. You might be a community's only link with the outside world when disaster strikes. You might overhear a weak distress signal from a foundering ship at sea and pass the word along to the authorities, all the while maintaining communications with the distressed vessel to reassure the anxious passengers and crew that help is on its way. You might be able to connect your ham station to your telephone line so that one of your fellow citizens can use his own phone to talk with a serviceman relative in some remote corner of the world through the facilities of your station. These are the glamour elements of ham radio, and they are commonplace enough so that they rarely make news any more.

But you don't need the glamour to have fun in ham radio. There's something of a touch of magic in having the capability of pressing a button and talking into a mike and having someone, perhaps many thousands of miles away, hear you and talk to you.

Many hobbyists prefer to talk to local stations only—people they make it a point to meet sooner or later. They use their stations as a kind of super intercom, keeping them set on a single frequency so they can hear all the goings-on of the members of their "group."

Amateur radio is a multifaceted world that can interest every type of individual. You may swap weather reports with kings, play chess with attorneys, talk "radio" to doctors by the hundreds, and hold idle conversations with bricklayers, barbers, garbage collectors, famous TV and movie actors, mentally retarded children, and even a Presidential candidate.

If there is one common denominator, it is the universal thrill of knowing your voice is being propagated through miles of air and space, to arrive intact at a destination no one can predict with certainty.

The overwhelming majority of individuals who become amateur radio operators never let their licenses expire. Once they've been won over to the exciting hobby, they begin to see the boundless areas of interest to explore. This book will serve to introduce you to some of the more popular aspects of hamming. The rest is up to you. This exciting world is out there waiting, and there are others just like you who are tuning their dial listening for your call.

# Introduction

BECAUSE YOU HAVE, BY PURCHASING THIS BOOK, IMPLIED that you are interested in, or at least curious about amateur radio, let me be among the first to extend a hand in welcome into this wonderful world. This book is intended to acquaint you with some of the many aspects of the hobby, and to get you started along the way to getting a license and becoming one of the family.

TAB has had this book in circulation for a long time. The first edition, written by Ken Sessions, came off the presses in 1974. But Amateur radio is a hobby that is very much alive, and in a constant state of change, so any book on the hobby that is to be of value to the reader must be frequently updated.

Several years ago, I found myself updating Ken's book, and this is my second rewrite of the book. To me, it is one of my more important writing jobs, due to the subject matter. I've been instrumental in bringing several new hams onto the air, and I hope that, by way of this book, I can help out a few more.

This edition includes some of the many changes made by the FCC since the previous edition was printed. More importantly, a few chapters have been added to give you more of the fundamentals you will need to get off on the right foot. Also, there is now some code-practice material so that, if you are studying on your own or with a buddy, you can still get some of that needed code.

The original material that remains from earlier editions has been gone over with a fine-toothed comb and extensively rewritten to make it easier for you to read and understand.

If you are indeed contemplating the idea of getting into ham radio, let me again welcome you into the hobby and, whether by way of this book or not, I hope you get as much out of the hobby as I have . . . and perhaps a little more.

# HOW TO BE A
# HAM

**3RD EDITION**

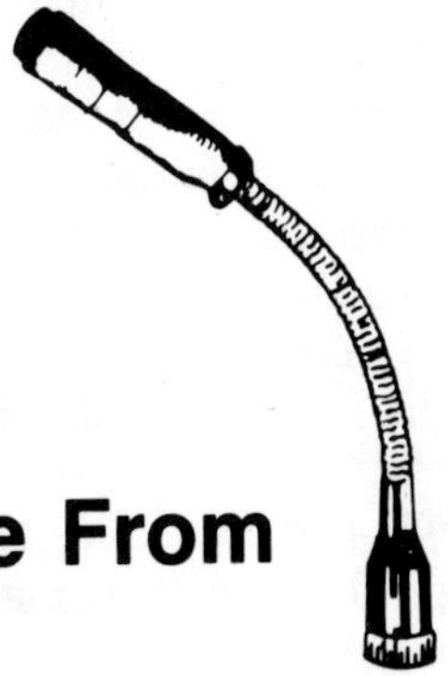

# Chapter 1

# What We Are and Where We Come From

As long as radio has existed, there have been independent experimenters who have wanted to try it for themselves. This is true for any art or science. Unlike people who practice the art for a living, experimenters do so out of an insatiable curiosity, or a love for the art. From the word *amor*, which means love, we get *amateur*, meaning one who loves. A radio amateur is simply one who loves the art of radio communication, and practices it for the fun of it.

Long before the advent of radio communications, there were people who went from place to place entertaining, simply because they enjoyed it. Since they had little to invest in costumes or makeup, they made their own. Makeup was made from a base of ham fat, and so they were called *hamfat actors*. This monicker was later shortened to ham actors, and the name has stuck to this day. When radio came along, the term was passed on to those who improvised their own equipment from whatever they could scrounge, and managed to get on the air on a non-commercial basis. From that day to this, we have been known as *hams*.

In the entertainment profession, a ham implies one with little talent. Not so in our hobby. We have had within our ranks pioneers in the art—Edwin Armstrong, to name one. In modern times, amateurs meet some of the finest engineers in the country. The beginner who passes his novice test today may well be professional a few years down the pike.

We, then, are the radio amateurs, also called hams. Some of us make our living in the radio communications or in the electronics industry. Others do not. When we fire up our rigs on the ham bands, however, we leave our jobs behind, operate for the pleasure of it, and assume an equal status with the rest of the hams.

What do amateurs do? They communicate. They discuss anything from the weather to politics and religion. They pass messages from one end of the hemisphere to the other. They experiment. They help in times of trouble, or with civic functions. Wherever communications via the air waves are involved, you will find amateurs.

You will find them playing chess. You will find them studying the Bible. You will find them bouncing signals off the moon. You will find them passing on a radiogram for some homesick GI. You will find them tracking satellites. Mostly, you will find them simply fellowshipping together and having a whale of a good time doing it.

Who are they? Anybody you can name—royalty, politicians, astronauts. There are young people, old timers, men, women, and kids. There are farmers, millionaires, and missionaries. Mostly, there are lots of ordinary people like yourself.

Neil Rapp of Vincennes, Indiana was only five years old when he passed his Novice license exam to become the world's youngest ham. That was in 1977. At an age when his peers were watching Sesame Street, he proved to the FCC that he was a competent operator. Since then, he has spent considerable time on the air just meeting other hams. Today (1986) he is active on 2 meters daily, passing message traffic (Fig. 1-1). He has been manager of the local 2-meter net, and edits the newsletter for the ham radio club in Vincennes. He is very active in ham radio, and never misses a hamfest or local event that involves ham radio. His interest in ham radio has opened up a bright future for him, and he is going full speed ahead to enjoy the greatest advantage he possibly can.

Jake Kaufmann is a Presbyterian minister, and an active ham. While he was pastoring the Brick Presbyterian Church in Perry, NY, he also spent much time on the air (Fig. 1-2). On one occasion, I watched as he made contact with a nun at the Mayo Clinic in Rochester, MN. With her was one of Jake's parishioners, a cancer patient. Through the medium of ham radio, a Catholic nun was helping a Presbyterian minister to visit one of his congregation! Jake has since moved on to a church in Wyandotte, Michigan, where he continues to use his hobby both for fun and as a tool for his work.

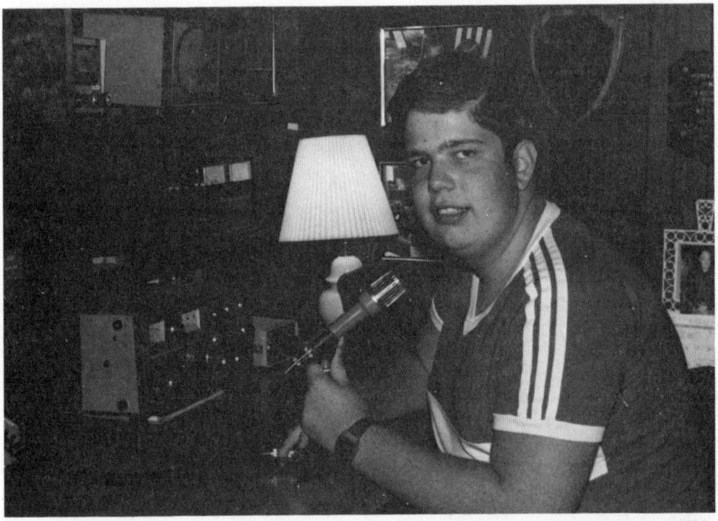

Fig. 1-1. Neil Rapp of Vincennes, Indiana obtained his Novice license in 1976, when he was five years old. Today, at 14, he has an Advanced license, and the memory of once having been the youngest ham in the world.

I was introduced to ham radio at the Bunker Hill Boys' Club in Boston, MA. Beginning with a teenage rivalry that led to a fight and a determination to prove that I could match my antagonist's accomplishments, I learned the code. Then a rivalry became a friendship as he and a kindly old man introduced me to ham radio. I don't mind saying that the rather shaky start there was my entrance into electronics as a profession. Thirty-three years later, I am still a ham, and make my living taking care of the electronics associated with the University of Rochester's nuclear accelerator. The build-it-yourself influence of ham radio has never left me: even the computer terminal on which this is being written is mostly homemade!

I could go on and on. There are many hams, and each has his or her own story to tell. Hams are simply a cross-section of the American people, bound together by our common interest in radio. You are welcome to become one of us.

In the early years of radio communication, anybody producing or operating radio apparatus was a picturesque individual. Radio was not like it is today. First of all, telegraph code was the only known means of putting information onto the air. Secondly, the only way radio signals could be produced was by means of a high-voltage electric spark.

The room where the apparatus was housed was an exciting place called the *shack*, and that, by the way, is where the name of a large chain of electronic parts stores originated. The shack was an exciting place with its humming transformers, and the loud, hissing, snapping spark. Amateurs in those days wound their own coils, and often built every bit of their gear.

As World War I came and went, the vacuum tube came into being. Radio equipment became more sophisticated. Voice communications became a reality. The number of radio stations on the air was so great that licensing and frequency allocations became necessary. Amateur operation was thought to be unimportant, so amateurs were restricted to wavelengths shorter than 200 meters, which were thought to be useless.

Necessity is the mother of invention, and the hams figured out the secrets of short waves. Successful contacts were made over greater and greater distances. The amateurs had proven themselves. Then came World War II. As the nation mobilized, it found

Fig. 1-2. The happy family of Jake Kaufman, W2GAT, with him in his radio room. He is a very active amateur, as you can tell by the many certificates of achievement on the wall. You'll find him on almost any band when he is not tending to the needs of his church.

a vast reservoir of competent radio operators needing little training.

By the end of the war, radio equipment had become well-developed. There was a glut of surplus equipment on the market which hams bought by the carload, modifying it for their own use. About that time, television came into its own as a public medium. The technology of TV equipment was very primitive, making it prone to interference from amateur equipment. For quite a while, many a ham found himself at odds with his neighbors, who were seldom sympathetic to his point of view.

Again amateur ingenuity won out, as the hams figured out how to improve their own equipment, and what steps should be taken with TV equipment. Nowadays, TV interference problems are comparatively rare.

Sometime between the wars, techniques were developed for producing a voice-modulated signal that was far more efficient and occupied only half as much of the radio spectrum as a conventional AM signal. This technique, called *suppressed carrier single sideband* (SSB), saw a lot of development after World War II, as the surplus market provided an inexpensive source of the necessary frequency control crystals. Today, SSB signals are used by most commercial and amateur voice modulated communications, other than FM stations. Even AM broadcasting stations use a modified version of this technique in which one sideband and a carrier signal is transmitted.

Today amateurs seldom build their own rigs. The complexity of the equipment, and the expensive test equipment needed to align it makes building from scratch impractical. However, there is much that the ham can and does build, and many at least know how to troubleshoot and repair their equipment.

Amateurs are deeply involved in satellite communications. A number of satellites, all named OSCAR (Orbiting Satellite Carrying Amateur Radio) have been orbited with the cooperation of the government, and many hams enjoy the experience of tracking and communicating with OSCAR.

Other hams work with super high frequencies, or with laser communications. As long as the art of radio communication has a frontier, hams will be on it.

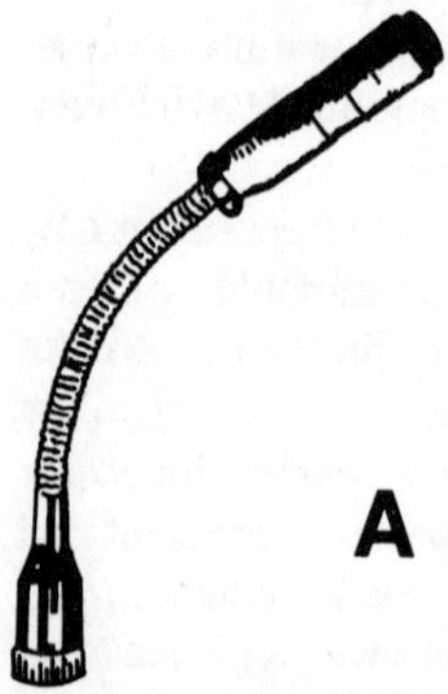

# Chapter 2
# A Reason to Be

AMATEUR RADIO DIDN'T HAPPEN BY ACCIDENT. FROM THE beginning of the science, we amateurs have earned our status. There are specific reasons cited in the FCC regulations for granting us this privilege. (It is a privilege, not a right.) These are several of the reasons for the existence of radio amateurs which are recognized worldwide.

## PUBLIC SERVICE

Part 97 of the FCC rules deals exclusively with Amateur Radio. The first entry in Part 97 specifically states that a fundamental purpose of that section of the Rules is to provide an amateur radio service that is of value to the public "as a voluntary, noncommercial service, particularly with respect to providing emergency communications."

This is perhaps the most important single justification for the existence of ham radio as an organized hobby—the potential good it can do a community, particularly in times of disaster or other emergency. The federal government recognizes the usefulness of radio at such times, and encourages amateurs to participate in skill-building programs designed to increase operator proficiency with regard to message handling, emergency team organization, and radio network setup and operation.

The instances where ham radio has paid off for the government are far too numerous to list in one small book. Let it suffice to say that, when a major disaster occurs, commercial communications systems can break down, either due to the same forces that caused the disaster, or from overload of message traffic. However well-designed a plan for disaster might be, it is still subjected to the unpredictability of human behavior. The best arrangements can, and do, break down. It is here that amateur radio comes into play as a backup system, and our service has often proved its worth in that respect.

Here's an example: a devastating earthquake levels a city. Power lines and telephone lines are down. Official emergency radio equipment is overloaded and perhaps partially out of commission. The result is an incredible log jam of communications that can bring even the best-planned system to a screeching halt.

But there are amateurs—people who seem to come out of the woodwork with an incredible array of portable and fixed radio gear. Often still wearing the bruises of the disaster themselves, they improvise, repair, and somehow come up with their own loosely-organized but efficient network that takes some of the load off the official emergency system, and things get moving again.

That is a hypothetical situation, but not by any means unrealistic, for time and time again, amateurs have been on the job when they were needed. They will continue to do so, never asking or accepting a penny for their services. That's the whole idea; that's why we're given a generous amount of space in what is known to be a finite, limited medium. We have always earned our privilege of using the airwaves, and we always will.

There are many true-to-life examples. When there were extensive, devastating forest fires along the California coast, many local residents weren't yet aware that a major disaster was taking shape. Amateur operators with mobile radios—radios with a far greater range and reliability than the best-designed CB could ever have—were assigned to help with evacuation from threatened areas. As the mobile units were deployed, they were kept informed of wind conditions, fire movement, and open routes for evacuation. In turn, as they ran into trouble—residents who refused to leave, new fire breakouts, etc.—they passed the word on to the net control at the local sheriff's substation. The result was that the evacuation was completed in record time, with a minimum of groping in the dark. In an emergency, communications can be the most important single commodity in existence. That's what we, as radio amateurs, have to offer.

Amateurs were on the scene when Mount St. Helens erupted in Washington State with losses of both life and property. They were there when tornados ravaged the Midwest. They were there to help when floods devastated large portions of several states, and when blizzards paralyzed the Northeast. Time and again they were there, and they will continue to be there.

On a more personal level, amateurs have provided the saving link in obtaining aid for persons injured in automobile accidents, finding people in downed aircraft, getting help to marine craft in trouble, finding lost kids, and providing extra communication for a community event. Emergency or otherwise, amateurs are handy people to have around.

When a disaster happens, the natural structure of amateur radio provides a world-wide network controlled by on-the-scene amateurs. It handles official and unofficial messages, incoming and outgoing, using its own standardized message format and priority policies. The usefulness of this voluntary service is virtually unlimited when it comes to providing emergency radio communications. With that in mind, it shouldn't seem too difficult to understand why hams are so often relied upon by the richest, best-equipped government in the world. Consider, for example, what happens when a dam collapses and torrents of water engulf residential and metropolitan areas. A small number of panic-stricken people can quickly tie up the telephone lines, preventing more vital communications from getting through—but the amateurs are there.

No sooner are the disaster victims aware of what is happening than those hams that are able to turn on their rigs. Within minutes they congregate about a previously agreed upon frequency, and exchange status reports. The local civil emergency headquarters, long aware of the value of amateurs, has an amateur station manned by volunteers on the air. While the populace is still staggering to their feet, the civil disaster headquarters is gathering a regional overview of the situation. Then, as the various civil agencies become coordinated, the amateur network takes up much of the overload in message traffic, leaving local government communications better able to function.

Somewhere a doctor is needed. An amateur assigned to the area passes the message to an amateur at or near the local hospital. There it is relayed to the hospital's resident communications center, and a doctor is on his way. Medications are needed. The request is passed by amateur radio to the pharmaceutical supply house and, with the cooperation of the police or the National Guard, they are on their way.

A power line is down, sparking its high-tension destruction. A passing ham flashes the word to disaster headquarters from his mobile rig, and help is on its way. And so it goes, until the services of the hams are no longer needed.

For many, such an experience might happen just once in a lifetime, as it has for me. You sure sleep well when you know you helped out, no matter how small the part. Now, how about the others, those who never meet with a need for their services? Of them, it is wisely said, "They also serve who stand and wait."

## TECHNOLOGICAL ADVANCEMENT

There is another reason why the government, in Part 97, sanctions our hobby. Since the art began, amateurs have contributed to the advance of the science. It was amateurs who first discovered the potential of the vast spectrum of wavelengths shorter than 200 meters. It was amateurs, with their tireless tinkering, that worked the bugs out of voice communication, radio teletype, slow-scan television (which can transmit pictures to earth from space satellites), and many other techniques. Nowhere but in the free world, where inquiring minds can seek out their own answers, can one find pioneer research at so low a cost. That doesn't mean you will be on the frontier of technology the moment you get your license, but you never know.

Thousands of people who work in research owe their start to amateur radio, myself included. There is no way of knowing how many ideas perfected in the lab had their beginnings in the imagination of a tinkerer who did his thing just for the joy of it.

## A RESERVOIR OF PEOPLE

Never were amateurs more appreciated by their government than at the start of World Wars I and II, when the country had to mobilize quickly in its own defense. The needed buildup of the armed forces necessitated training personnel to handle communications. This training was reduced by the discovery of a vast number of men and women who were already familiar with radio communications techniques, and had only to be taught the official military procedures. As long as the hobby exists, we will be there when we're needed. The government recognizes this in Part 97, which lists "expansion of the existing reservoir within the amateur radio service of trained operators, technicians, and electronics experts" as part of the purpose of amateur radio.

## GOODWILL AMBASSADORSHIP

Finally, the government regulations cite the enhancement of international goodwill. No amount of propaganda or diplomacy can match the benefits of one-to-one contact between amateurs of different countries. They quickly learn that other hams are people with similar needs and interests, and the suspicion born of isolation melts away. (Remember this the first time you talk through the Iron Curtain.)

Citizens of some foreign "cold-but-friendly" countries might be bombarded with news telling how terrible things are in this country. We know some governments sift through the news and allow only the negative side of the U.S. to reach their people. As a result, we are thought of as the "ugly Americans"; or as an oppressed lot, brutalized by a murdering government, with no recourse whatever; or as an insensitive, money-hungry lot out of tune with the rest of the world. Communication makes the difference, and shows the lies for what they are.

Radio amateurs are far more than a number of persons pointlessly playing with the air waves. We exist for a purpose, and share the common goal of perfecting our abilities to maintain, operate, and understand our radio equipment. If you see a value in these objectives, and love the communications art, we welcome you among us.

# Chapter 3
# The Licenses and the Action

CERTAIN PARTS OF THE RADIO SPECTRUM ARE ALLOCATED specifically to amateurs, and they are allowed to operate only on those authorized frequencies. Moreover, the licenses given for amateur operation are set off into five specific classes: Novice, Technician, General, Advanced, and Extra. The lower classes are more restricted in the frequencies and modes of operation than are the higher classes. This is not to imply that any kind of snobbery exists; while you may find a weirdo here and there who can't remember when he was a beginner, ham camaraderie generally extends between license classes.

## THE NOVICE

The Novice license is the bottom-grade, easiest-to-get amateur license. Consequently, just about everybody starts out as a Novice. A Novice is limited to the transmission of code, rather than voice, and is restricted to a few portions of the amateur bands. This, however, is no reason to think the enjoyment of ham radio is limited; in fact, your operating speed increases the quickest when you're actually using the code to enjoy yourself. By making this license class available, the government has provided an easy way for you to work your way up to the proficiency required for the higher grades.

Code is fun. In fact, many amateurs continue to operate in code long after graduating to the higher grades. Large portions of many of the bands are reserved exclusively for code (CW) operation.

A Novice must limit his or her transmitting power to no more than 200 watts peak envelope power. This makes sense when you realize that a Novice operator is, after all, very new to the hobby and, as such, will make mistakes. Limiting the transmitting power limits the amount of havoc a beginner's mistakes will cause. The license is good for ten years, and you can renew it after that time. This is a far cry from the first licenses issued to Novices back in the early fifties. They were limited to 75 watts, crystal control, and the license was good for only one year with no renewal.

Novices may operate in the following frequency bands:

| | |
|---|---|
| 3.7 to 3.75 MHz. | (80 meters) |
| 7.1 to 7.15 MHz. | (40 meters) |
| 21.1 to 21.2 MHz. | (15 meters) |
| 28.1 to 28.2 MHz. | (10 meters) |

The Novice license is the introductory license, and the study material covers the rules you need to know in order to stay out of trouble, plus a few basic ideas of how radio signals behave as they zip around the world. Holding a Novice license provides you with a means of learning as you go.

## THE GENERAL AND TECHNICIAN

The General class license is the reward you get for additional learning. The exam requires a bit more knowledge of radio communications theory and law, which is fine because by now you'll be a little more curious as to how some of these circuits work, and you'll be using more sophisticated equipment that requires greater understanding. For example, since you may be operating on many different bands, you'll need to know more about antennas and how they work.

In contrast to the somewhat restricted nature of the Novice license, the General class permits a very wide range of operating privileges on frequencies ranging from just above the AM broadcast band up to the very, very short waves known as *microwaves*. Furthermore, a General-class licensee can use voice communications and, on segments of these bands, such exotic modes as slow-scan television, radioteletype, and satellite communications.

The General class license exam requires a code operating speed of 13 words per minute. The license is good for ten years, and is renewable.

The Technician class license requires a code operating speed of only 5 words per minute, the same as the Novice. The written test is the same as for the General class, and the operating privileges consist of some of both worlds. A Technician licensee has all the operating privileges on the lower-frequency bands that a Novice has, and he can operate with General privileges on the VHF bands. The Technician license is intended for the experimenter whose prime interest is voice operation and the development of the art in very high frequencies, which were once thought useless.

## THE ADVANCED AND EXTRA

While the General license permits an amateur to sample almost everything going on in the radio world, there are many who like to dig deeper into the mysteries of radio and electronics. The Advanced and Extra classes are designed for these amateurs.

Both of these license exams require a greater knowledge of theory. You'll dig into such things as why transistors work, what is currently being done in integrated circuits, how a television picture is transmitted, logic circuits and definitions (as in computers), and much, much more.

The Advanced license doesn't require any faster code, but it does demand a greater knowledge of theory. In return for that, you get additional operating privileges on some portions of the bands that General licensees do not have.

The Extra exam is harder still. It requires a code operating speed of 20 words per minute, sending and receiving, and a written exam comparable to that required for commercial radio licenses. It is the "top of the line" in amateur licenses, and many consider it a prestige license. Those who pass the Extra exam can receive a license certificate that is suitable for framing, in addition to the wallet-sized card that comprises the license for the other classes.

## THE ACTION

"Once I get the license and I'm on the air," you may ask, "what then? What do hams do?" I could fill a whole book with the answers to that question. First of all, we visit with one another over the air. Then, as we get to know one another and discover what we have in common, we use ham radio as a medium in which to get together and do our particular thing.

On a one-to-one basis, you can discover that the person at the other end is perhaps a boating enthusiast, a stamp collector, is into radio-controlled models, builds and operates model trains, races sports cars, is a photographer, raises exotic flowers, keeps bees, hikes, camps, jogs, or does any of the hundreds more interesting things people do.

You meet the most interesting people over the air. You could find yourself talking with Senator Barry Goldwater, K7UGA, or perhaps with Tom Christian, VR6TC, a descendant of the Bounty mutineers living on Pitcairn Island. You could even find yourself talking with royalty—King Hussein of Jordan has the amateur call-sign JY1.

## NETS

When a number of hams with a particular interest get together, they often form what is called a *net*. A net is a sort of club meeting held on the air. There are a great many nets, each with its own unique flavor. Some get together for the sole purpose of passing messages along for other people, called *handling traffic*. You can send a radiogram anywhere in the western hemisphere without cost by way of traffic nets.

Then there are nets that meet on the air for everything ranging from Bible study to playing chess to emergency preparedness. There are nets of war veterans sharing their experiences. There are Masonic nets, missionary nets, antique radio nets—you name it, it's there.

Sailing enthusiasts often use amateur radio as a low-cost substitute for high-frequency maritime communication. They have their own nets, each covering its own particular part of the coast. There are also nets for balloonists and RV enthusiasts.

## AMATEUR TELEVISION

Some hams have systems for sending *slow-scan* television. Slow-scan is a means of sending still pictures by radio. Since it takes a much longer time to send a single frame by slow scan than it does with conventional television, and since less detail is sent, a picture can be sent around the world within the limits of a voice communications channel. It can even be stored on tape with an ordinary cassette recorder!

Conventional television is called *fast-scan*. Amateurs are into that, too. Most of the larger cities have amateur television (ATV) clubs, and many have a *repeater,* which receives a signal from a ham anywhere in town, amplifies it and retransmits it with enough power that it can be received as much as 75 or 100 miles away. This sort of operation goes on in the 420 MHz amateur band. If your own TV set could receive the amateur TV signals, they would be found just below the UHF channel 14. In fact, many hams modify TV sets just to do that!

## 2-METER REPEATER OPERATION

Perhaps you like the idea of a repeater, but you're not really keen on getting into fast-scan television. Well, there are repeaters for voice communication. One of the most popular bands for this is 2-meters (144 to 148 MHz). On this band, which can be picked up with any radio that covers the public service frequencies in that region of the spectrum, the emphasis is on portability. Some hams have rigs in their cars; others use hand-held walkie-talkies. With the repeater, they can chat with a person 100 miles away while walking down Main Street. In addition, many repeaters have connections into the local telephone system. Hams who have key pads on their radios similar to the key pad on a telephone, can dial any number in the city through the repeater. Connection between ham equipment and the telephone system is called *phone patch*.

2-meter repeater operation is an important link between ham radio and the public. It provides highly reliable local communication, and hams often use it to assist with community events. They also use it in emergencies, whether large or small. There has been more than one instance where a timely phone call to the police, made from a ham in his car or on foot by way of the repeater, has helped save a life.

## MARS

MARS is a very special extension of ham radio. It operates on frequencies outside the ham bands, but its members are hams first. The letters stand for Military Affiliate Radio Service, and MARS is a concordant service involving many amateurs. In this service, amateurs volunteer their equipment part of the time to become a part of the military communications system. MARS operation consists mainly of handling messages on behalf of the men and women in the military service. Member stations use call letters assigned by the particular branch of the military they are affiliated with, and the operation is carried on using frequencies outside the regular amateur bands.

MARS operation is sponsored by the Army, the Air Force, and

Fig. 3-1. A Field Day adventure with your fellow hams can be an unforgettable experience (photo by WA2JHD).

by the Navy/Marine Corps. Each has its own network of affiliated stations, and each has its own unique operating procedure.

## MESSAGE TRAFFIC

There are many third-party message nets which operate with

Fig. 3-2. Civic authorities hold amateur radio in such high regard that they lend their facilities to the local club as a meeting place (courtesy South Towns Amateur Radio Society).

Fig. 3-3. Club room of South Towns Amateur Radio Society (photo courtesy of South Towns Amateur Radio Society).

telegraph code. Others use voice communication. Still others use *RTTY* (radio teletype). Much RTTY work is still done with mechanical machines, but the FCC now allows the use of the ASCII computer code; now many hams talk to each other computer-to-computer.

Imagine sending a message from one end of the country to the other in just a second or two, including an acknowledgment of receipt. Or, how about a game of Space War with a ham and his computer in Alaska, Canada, or halfway around the world?

Need a formula to design a new antenna you want to make? Need design information for a filter to make your receiver work better? Punch into a computer net and see if anyone has what you need. If so, it can be sent to your screen in a fraction of a second, and your computer can print it out on paper for a permanent record if you decide you like it!

Are you interested in high technology? Amateurs have repeaters in orbiting satellites. Others communicate by bouncing signals off the moon. Still others experiment with laser communication. What can you do once you're a ham? Just name it! Your ham equipment can become an extension of yourself, giving you global coverage under emergency conditions (Fig. 3-1), from your home, or from a club station (Figs. 3-2 and 3-3).

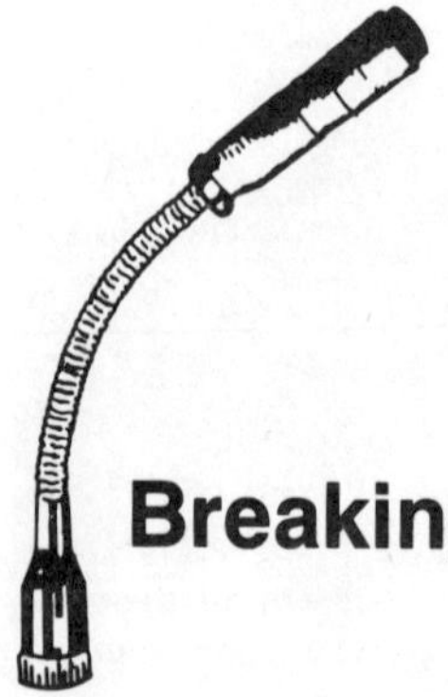

# Chapter 4
# Breaking Into Amateur Radio

IF YOU LIKE THE IDEA OF BECOMING A HAM RADIO OPERATOR, the first question to ask is how to go about it. You are aware that, to become a ham radio operator, you must know the Morse code, and you must learn enough radio theory to pass a government examination. But how well must you know the code, and how difficult is the examination?

If you've had no previous contact with radio communications before, ease your way into a General class license by way of the Novice class. That's what the Novice class is for.

The tests are not all that difficult; still, it can be rather unnerving entering a heretofore unknown situation. You are not alone. I know of no radio amateurs that got their start without the help of another amateur. Wherever in the country you may be, you need only contact the nearest amateur radio club, and you will have ample help. If a local amateur is unable to find the time to help you himself, he or she will almost certainly direct you to the nearest club.

It looks as if the government has recognized the spirit with which hams help one another into the hobby. License examinations were originally given at local FCC field offices. Now they are given by volunteer examiners, themselves hams, who have gone the same route as you are going. Every local club either has organized licensing classes, staffed by hams, or has hams willing to work with you on a one-to-one basis. If you cannot find the local club, the local

Civil Defense or Red Cross office can probably help you. If not, the local FCC office can give you the name and address of the local Volunteer Examiner Coordinator who will in turn direct you to help. Alternatively, you can write to any of the several amateur radio magazines, or to the American Radio Relay League at 225 Main Street, Newington, CT 06111. No, indeed, you are not alone.

## LEARNING THE RADIOTELEGRAPH CODE

There are two steps to learning the code. First, memorize the sequences that represent the various alphanumeric characters used in CW communication. Then practice until you can immediately recognize any letter you hear. It is best if you can work with a group of other beginners practicing the code and studying for the exam. This can be done either in a formalized class run by a local club, or with one or two others under the supervision of a friendly neighborhood ham. See Table 4-1.

There are a great many recorded code-practice courses available. Some are quite good; others are not. With any course, beware of the possibility of memorizing the material. This gives you the illusion of a much higher reception speed than you may actually have.

One of the best methods of code practice, outside of a class, is the buddy system. Find another person who will work with you, sending code for you to copy, or listening to the code you send. When you practice this way, the material can be mixed up so that it can't possibly be memorized.

## CODE PRACTICE MATERIAL

The ARRL has recorded code courses and has published books to assist in the instruction of code to groups. Over the years they have used various formulas and sets of study material. It is still considered by many the most effective code study material available. It consists of groups of letters giving you a portion of the alphabet at a time. With each group of letters there are words, phrases, and sentences.

This particular material was designed to provide practice sending the code, although I have used it both for sending and receiving. It can be used most effectively with the assistance of a person who already knows the code. This person, your instructor, should send you practice from one of the groups of letters until you know it. Begin at a very slow speed and work your way up.

**Table 4-1. The Morse Code.**

A . ___
B ___ . . .
C ___ . ___ .
D ___ . .
E .
F . . ___ .
G ___ ___ .
H . . . .
I . .
J . ___ ___ ___
K ___ . ___ (Go ahead)
L . ___ . .
M ___ ___
N ___ .
O ___ ___ ___
P . ___ ___ .
Q ___ ___ . ___
R . ___ . (Message received)
S . . .
T ___
U . . ___
V . . . ___
W . ___ ___
X ___ . . ___
Y ___ . ___ ___
Z ___ ___ . .

1 . ___ ___ ___ ___
2 . . ___ ___ ___
3 . . . ___ ___
4 . . . . ___
5 . . . . .
6 ___ . . . .
7 ___ ___ . . .
8 ___ ___ ___ . .
9 ___ ___ ___ ___ .
0 ___ ___ ___ ___ ___ or _________
Period . ___ . ___ . ___
? or "Repeat" . . ___ ___ . .
& ($\overline{\text{ES}}$) . . . .
Comma ___ ___ . . ___ ___
"Hey!!" ___ . ___ . ___

Ends of message (AR) . ___ . ___ .
End of operation (SK) . . . ___ . ___
One moment (AS) . ___ . . .

Fraction bar ___ . . ___ .
"Oops!" (error) . . . . . . . .
"E-r-r-r . . . u-m-m" (Double dash) ___ . . . ___

This practice material is from the ARRL publication "Learning the Radiotelegraph Code," and is reproduced here with permission from the ARRL.

## Group 1

Characters used: E T A O N I S

| | | | | | |
|---|---|---|---|---|---|
| AN | NO | ON | IT | TO | SO |
| IS | OAT | TAN | TEN | ONE | ANT |
| NIT | TOT | NET | TIE | ETA | TIN |
| TOE | SON | SIN | SIT | SET | ASS |
| SEA | INTO | TENT | TONE | ANON | NINE |
| NEAT | TOTE | TOOT | NONE | TEAT | TINT |
| ETON | SEAT | EASE | SENT | TAINT | STINT |
| EATEN | STONE | TOAST | TEASE | SNOOT | SENSE |

a) SIN NO ONE
b) IT SENSES TASTE
c) NONE IS SENT
d) SNOOTS EATS OATS
e) NO SEASON IS SET
f) IT IS NOT TEA
g) ETA IS A NOTE
h) NO SENSE IN EASE
i) STAINS TINT A SEA
j) ASSES NET NO SENSE

k) SONNIE SAT ON A TIN SEAT
l) IT TOASTS IN TEN TASTES
m) AN OASIS IS NEAT EASE
n) TIE A NINETEEN TON STONE TO A TENT
o) TOAST IS EATEN AT ETON
p) STAINS TAINT ONES NEATNESS
q) TOTE A STONE TO TENNESSEE

| | | | | |
|---|---|---|---|---|
| ETO | ANI | SIN | AOT | ESI |
| SAOT | NITN | SOOE | IATE | ANOE |
| SINES | TOEIS | NAOTE | ONTIS | ESTAO |
| ENTOIS | TOOSEI | OENTST | ANSENO | NANOSE |

## Group 2

New characters used: R H D U C M L

| | | | | | | |
|---|---|---|---|---|---|---|
| RAT | RUN | RUT | RICE | RUSH | RASH | RADIO |
| HAD | HUM | HAM | HAND | HELM | HULL | HOUSE |
| DAD | DAM | DON | DUDE | DICE | DOLT | DUNCE |
| COT | COD | CAM | CHIT | CORD | CURT | CHORD |
| MAN | MAD | MET | MOSS | MICE | MODE | MODEL |

| | | | | | | |
|---|---|---|---|---|---|---|
| LAD | LET | LID | LEAD | LORE | LUST | LURID |

| | | | | |
|---|---|---|---|---|
| REDUCE | RADIUS | RANCID | RADICAL | MARCONI |
| HALTER | HUDDLE | HORROR | HUMIDOR | ULTIMATE |
| ULSTER | UNLACE | URCHIN | UNARMED | STATION |
| DOLLAR | DISCUS | DANCER | DUCTILE | RESISTOR |
| COLLAR | CRUISE | CUDDLE | COLLIER | ANTENNA |
| MUDDLE | MURMUR | MOTHER | MIRACLE | AERIAL |
| LEADER | LANCER | LUSTRE | LECTURE | COIL |

| | | |
|---|---|---|
| TRANSMITTER | CHROMOSOME | MEDITERRANEAN |
| RADIATION | DASTARDLINESS | NECESSITARIANISM |
| OCCLUDED | CHROMIUM | OMNISCIENCE |
| ANTHRACITE | EDUCATIONAL | RESIDENTIAL |
| AMERICAN | THERMITE | SANITARIUM |

a) HAMS ARE ACTORS TOO
b) MIRACLES SELDOM OCCUR
c) MUSIC HATH LOTS TO CHARM
d) MAD CATS MURDER MICE
e) SEND CODE AND SUCCEED
f) SOME DANCERS ARE CRUDE
g) HENS HATCH SMALL ROOSTERS
h) MORSE CREATED THIS MIRACLE
i) CURSES ON ALL HEARSES
j) RURAL HOUSES ARE HUMID
k) ANTHRACITE COAL IS HARD
l) RESISTANCE IS MEASURED IN OHMS
m) UNCLE SAM NEEDS US ALL
n) THE SUN SHINES IN CONNECTICUT
o) CODE CLASSES ARE LITTLE TOIL
p) CHARACTERS ARE DOTS AND DASHES
q) CODE IS HEARD NOT SEEN
r) MARCONI LECTURED ON CURRENT RADIATION
s) INTO DEATH MARCHED THE THREE HUNDRED
t) AMATEUR RADIO IS AS OLD AS THE RADIO ART

| | | | | |
|---|---|---|---|---|
| URT | HLU | ADC | CIM | SLE |
| LRTM | OAUD | NICM | SRHL | MAUE |
| HSDNM | RHUAC | ODNCL | SIDCO | IUHER |
| TCODME | ADHRHU | CNDUTR | SLMIRU | LSEIRC |

## Group 3

New characters used: P F W Y G B

| | | | | | | | |
|---|---|---|---|---|---|---|---|
| FOW | PEP | PIG | PAGE | PYRE | PITY | POPPY | PUFFY |
| POP | FRY | FIG | FROG | FLOW | FILE | FABLE | FOGGY |
| WEB | WON | WHY | WHIG | WOLF | WING | WHIFF | WEIGH |
| YIP | YOU | YEG | YOWL | YELP | YAWN | YOUNG | YACHT |
| GOB | GYP | GIN | GLOW | GASP | GRUB | GABLE | GLOBE |
| BOY | BEG | BIT | BANG | BEEF | BLOB | BADGE | BEFOG |

| | | | | |
|---|---|---|---|---|
| PACIFY | PEBBLE | PHLEGM | PEPPERY | AFFLICT |
| FEEBLE | FIDGET | FLINTY | FAIRWAY | BABYLON |
| WAFFLE | WEAPON | WINDOW | WAYWARD | CABBAGE |

| | | |
|---|---|---|
| DEPOSITORY | ELECTROTYPE | FISHWIFE |
| PYGMY | GUYWIRE | TAPE |
| ALPHABET | PRACTICE | MESSAGE |
| SUPERHET | GEAR | SIGNAL |
| CAPACITY | FLYWHEEL | FILTER |
| CRYSTAL | PURPOSE | STABILITY |
| FACTOR | BEATNOTE | AMPLIFIER |

a) EPIGRAMS SAY MUCH IN FEW WORDS
b) WHAT IS WORTH DOING IS WORTH DOING WELL
c) FORTUNE BEFRIENDS THE BOLD
d) ILL BLOWS THE WIND WHICH PROFITS NOBODY
e) WINGS FLY FAST
f) WHIFFS OF FOG PACIFY FEEBLE AFFLICTIONS
g) BOYS YELP WHEN BIT BY DOGS
h) A PUFFY BLIMP IS A PEPPERY FISHWIFE
i) TETRODE AMPLIFIERS RELAY POWERFUL SIGNALS
j) RUBBER WHEELS AFFECT RIDING COMFORT
k) PUT YOUR FLYWHEEL IN GEAR
l) GASP AND YOWL BUT BUY DEFENSE BONDS
m) WHOOP IT UP BEFORE DYING
n) LIGHT GLOBES GLOW WITH POWER
o) TIGHTEN CAP BEFORE TIPPING
p) THE BADGE OF COURAGE IS WON BY PLUGGING
q) ALWAYS COPY BEHIND IF POSSIBLE

r) BICYCLE RIDING CAUSES PUFFING
s) COPYING BY TYPEWRITER IS GOOD PRACTICE
t) PYGMY GUYWIRES PUT STRENGTH INTO POLES

| | | | | |
|---|---|---|---|---|
| HFM | ODW | AUY | NCG | IMB |
| FHEP | PTRF | YOUW | WDAY | BMNG |
| BCADF | GUOHP | YDTRB | WHEMG | FRSCY |
| PSDGCB | FUIYW | MAWPFLY | GHORGB | WRNFEY |

**Group 4**

New characters used: J K Q X Z V

| | | | | | | |
|---|---|---|---|---|---|---|
| JET | JUNK | JAVA | JERK | JACOB | JAPAN | JELLY |
| ASK | KINK | KEEP | KALE | KHAKI | KNACK | KNAVE |

| | | | | | | |
|---|---|---|---|---|---|---|
| QUO | QUIT | AQUA | QUIZ | QUICK | QUAIL | QUEEN |
| JIM | OXEN | AXLE | XRAY | EXACT | AXIOM | OXIDE |
| ZOO | ZERO | JAZZ | ZING | AZURE | CRAZY | ZEBRA |
| IVY | VINE | HAVE | VOID | VIVID | AVAIL | EVERY |

| | | | | |
|---|---|---|---|---|
| JACKAL | JERKIN | JOVIAL | JONQUIL | ZENITH |
| KIBITZ | KINGLY | KAISER | KICKOFF | EMBEZZLE |
| QUARTZ | QUAVER | QUORUM | ACQUIRE | PULVERIZE |
| PICKAX | EXCUSE | XERXES | EXPLAIN | GRAZE |
| ZEPHYR | ZIGZAG | ZOUNDS | BUZZARD | OXYGEN |

a) JACOB ASKED THE KING FOR QUARTZ OXIDE
b) QUICK AXIOMS AVOID UNJUST TAXES
c) QUIZZES QUICKEN THE REFLEXES
d) JOVIAL KIBITZERS ARE VERY OBNOXIOUS
e) JAPANS JUNK QUITS QUICKLY
f) EVERY QUEEN KEEPS JONQUILS
g) HEXAGONAL KNUCKLES HAVE CRAZY KINKS
h) XERXES EXPLAINED THE QUICK VICTORY
i) VELVET QUAIL GRAZE IN ZOOS
j) KINGLY BUZZARDS QUAVER IN AZURE SKIES

| | | | |
|---|---|---|---|
| JPR | SNQV | QYLGV | ZKJFBG |
| FKH | JAHK | XWOFZ | RHKPVO |
| DQW | ZGVM | KPCAJ | QLFPDQ |
| XYU | YXUT | BZLNX | XMZNJU |

| | | | |
|---|---|---|---|
| ZGM | ZWDO | IVKQW | YKWZBG |
| BVC | KBLI | YJCTV | HSIXJA |
| QPX | JFRE | EQSXR | MCVXZQ |

### Group 5

Beginning with this lesson, the instructor may make good use of newspaper and other text for practice material.

New characters used: 1 2 3 4 5

| | | | |
|---|---|---|---|
| 215 | 3242 | 34125 | 244521 |
| 431 | 2151 | 42153 | 352142 |
| 234 | 4531 | 33214 | 332312 |
| 551 | 5324 | 24115 | 545134 |
| 422 | 1543 | 53425 | 152342 |
| 342 | 3154 | 14523 | 453152 |
| 351 | 2231 | 21435 | 514324 |

| | | | |
|---|---|---|---|
| AE1 | CP2X | 3Z4BJ | 123VJQ |
| 3T5 | 14B3 | XYZ51 | W3X5AM |
| 420 | F3G4 | 3KP14 | 435PT1 |
| N43 | EAR5 | RY311 | AW1JOZ |
| 2MC | 424V | 14V4U | 2UV4JB |
| HT5 | D1J3 | L3VJ4 | XU4PY1 |
| 425 | 412V | 21JJ2 | 5SH32W |

### Group 6

New characters: 6 7 8 9 0

| | | | |
|---|---|---|---|
| 867 | 9768 | 77689 | 967760 |
| 906 | 0069 | 90870 | 806970 |
| 760 | 7987 | 68096 | 688979 |
| 896 | 0869 | 86970 | 087068 |
| 707 | 8776 | 09906 | 786096 |
| 998 | 8609 | 87780 | 779680 |
| 807 | 7068 | 87669 | 696087 |
| 196 | 3874 | 62840 | 647359 |
| 245 | 1928 | 17395 | 821073 |
| 837 | 5603 | 61723 | 489625 |
| 604 | 7495 | 89450 | 107446 |

| | | | |
|---|---|---|---|
| 932 | 1620 | 42417 | 829310 |
| 758 | 4835 | 38950 | 593758 |
| 103 | 2071 | 59636 | 621054 |

| | | | | | |
|---|---|---|---|---|---|
| C3B7 | Q78G | F3M0J | Y9UBC | A1B3C2 | 9105P4 |
| JW19 | H5SH | 132R5 | WJV41 | 4D6F5E | Q6U3V4 |
| 47BE | 4CY3 | 9L4PV | 0OX23 | 9HG7J1 | SM3801 |
| X6Z2 | ZK67 | K176B | PQ954 | 8L0KM3 | 6Q7GT |

**Group 7**

By now most students should be able to recognize letters and numerals immediately when heard. If not, more practice is indicated, especially on the characters causing trouble. It can be helpful to send letter pairs frequently confusing to the beginner, such as G-W, F-L, O-S, P-X, Y-Q.

| | | | | |
|---|---|---|---|---|
| GOWN | FILM | ARK | JAW | OASIS |
| APEX | QUERY | WAGER | LEAFLET | WORKER |
| | BLOSSOM | | EXPERT | |

The following sentences employ every letter of the alphabet.

a) THE EXPLORER WAS FROZEN IN HIS BIG KAYAK JUST AFTER MAKING QUEER DISCOVERIES
b) WHENEVER THE BIG BLACK FOX JUMPED THE SQUIRREL GAZED SUSPICIOUSLY
c) MY HELP SQUEEZED IN AND JOINED THE WEAVERS AGAIN BEFORE SIX OCLOCK
d) WE DISLIKE TO EXCHANGE JOB LOTS OF SIZES VARYING FROM A QUARTER UP
e) A QUART JAR OF OIL MIXED WITH GOOD ZINC OXIDE MAKES A VERY BRIGHT PAINT
f) WHILE MAKING DEEP EXCAVATIONS WE FOUND SOME QUAINT BRONZE JEWELRY
g) SIX JAVELINS THROWN BY THE QUICK SAVAGES WHIZZED FORTY PACES BEYOND THE MARK
h) THE PUBLIC WAS AMAZED TO VIEW THE QUICKNESS AND DEXTERITY OF THE JUGGLER
i) WE QUICKLY SEIZED THE BLACK AXLE AND JUST

SAVED IT FROM GOING PAST HIM
j) THE JOB REQUIRES EXTRA PLUCK AND ZEAL FROM EVERY YOUNG WAGE EARNER

### Group 8

New characters used: comma, period, interrogation, double dash, end of communication, wait, fraction bar, end of message.

Practice material from here on consists of plain text, now including the punctuation indicated here. Other punctuation won't be necessary, as it is not customarily used in amateur operation.

To acquaint you with receiving conditions, as you might find them in actual communication, play a tape from a communications receiver, or tune one in live while practicing. Decrease the volume of the practice tone source. Use actual radio communication for practice material. (The preceding code practice material is adapted from *Learning the Radiotelegraph Code*, Twelfth Edition, now out of print, Copyright 1970 by The American Radio Relay League. Used with permission.)

## THE WRITTEN TEST

The written examination is put together by an FCC-appointed volunteer examiner coordinator. He selects the questions from a set of standard questions for the particular exam level, which the FCC supplies. The standard question sets are included in the Appendix. There are two ways to acquire the necessary knowledge to answer the questions: You can attend classes sponsored by a local amateur club (which is probably best), or you can study on your own.

Chapters 8, 9, and 10 include some general information of several aspects of radio communication. If you have no prior knowledge of electricity whatever, try reading TAB #1507, *Beginner's Guide to Electricity & Electrical Phenomena*. With the fundamentals carried in that book and what is presented here, you should be able to make the grade.

## SETTING UP

Unless you have money to burn, it is best to start with a modest setup. In this game, there is absolutely nothing wrong with starting with used equipment, if you can't afford to buy it new. A local hamfest is an excellent place to find used equipment, but you should have an experienced ham with you to advise you on what to buy and what to pay for it.

Then there are those who prefer to buy a kit and assemble it themselves. That's fine, but unless you are very meticulous in your hand work, you can end up with a can of worms. More than one ham has just about ruined a perfectly good rig simply through assembly and soldering inexperience. If you have to send the assembled rig back to be debugged, it can get expensive. Whether you build a kit, buy used, or buy new, don't get too elaborate until you know you will be in the hobby for good. Let your equipment accumulate as you grow in experience (Figs. 4-1, 4-2, and 4-3).

There are still a great many hams that use a separate transmitter and receiver. There's nothing wrong with that, but setting up can be much easier for a beginner if you use a transceiver. With both functions combined in a single unit, there is less that can go wrong in the initial setup.

The minimum equipment for setting up a station includes transceiver, SWR meter, low-pass filter, lightning arrester, ground rod, and antenna. The antenna will be discussed in detail in a later chapter.

Figure 4-4 shows the connections of a typical setup. For the station ground terminal, heavy conductors run from the equipment to the ground terminal. If more than one piece of equipment is involved, each should have its own wire to the ground terminal, and

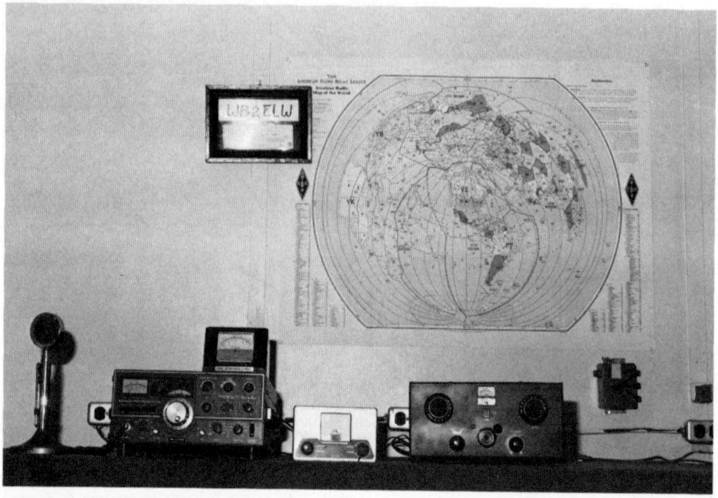


Fig. 4-1. It doesn't take a lot of expensive equipment to set up a neat but effective station (photo courtesy South Towns Amateur Radio Society).

Fig. 4-2. Your station can grow as you gain experience, as this impressive array shows (photo courtesy WB2HCT).

Fig. 4-3. All some hams need is a corner in their home (photo courtesy WA2JHD).

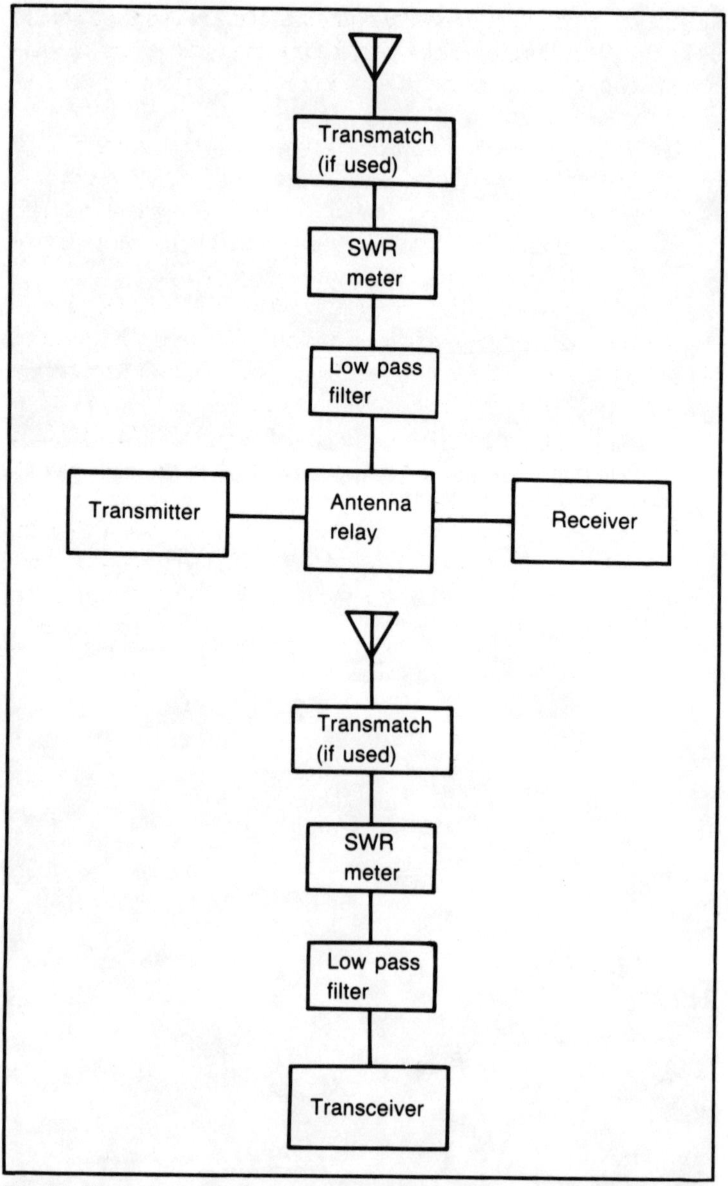


Fig. 4-4. Block diagrams of typical amateur stations. The top shows how a separate transmitter and receiver are connected. A transceiver is shown on the bottom. The use of a transmatch, or antenna tuner, depends on what kind of antenna you have. While you can send and receive without the low-pass filter or the SWR meter, their use is strongly recommended. Ground connections, not shown, should consist of an individual wire from each piece of equipment to the common ground terminal of the station.

all ground conductors from the equipment to the terminal should be of the same length. Run a very heavy conductor from the station ground terminal by the most direct route possible to the earth ground connection.

The earth ground connection consists of one or more ground rods. The number of rods depends on the conductivity of the soil. Dry, sandy soil is the poorest conductor, while if you live in a salt marsh, you couldn't do better! The arrangement shown in the illustration is for relatively non-conductive ground.

A low-pass filter is mandatory. Don't be afraid to invest in one. It can save you a lot of headaches. Be sure the design impedance of the filter matches the characteristic impedance of the transmission line; otherwise it won't work at its best.

Lightning protection is always a good idea, and it is a must if you live in an open area. While the degree of protection needed can vary with locality (equipment that is overprotected on the California coast will go up like a roman candle in Kansas or Florida) you should never be without *some* protection. A coaxial lightning arrester is cheap compared to the cost of your gear. Again, it's a good idea to get the advice of several local hams before you begin.

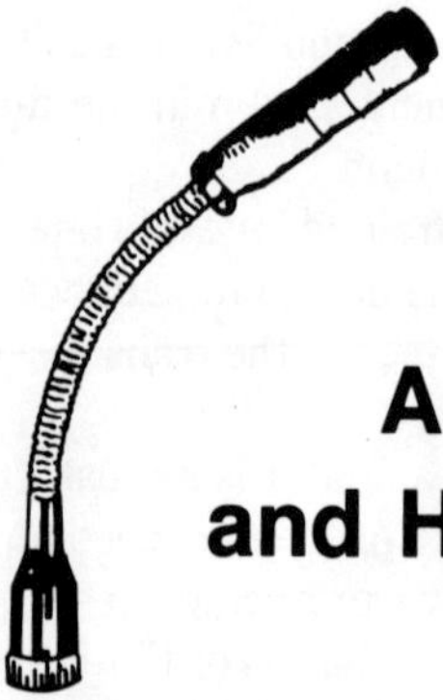

# Chapter 5
# Amateur Bands and How They Behave

PERHAPS THE MOST DIFFICULT CONCEPT TO UNDERSTAND in amateur radio is that of frequency. Is it a wavelength? A place on the dial? Is it, as the word implies, a rate of repetition? In a way, it is all of these. A radio wave, if it could be seen, might resemble an ocean wave. It has peaks and troughs, and it travels away from its source. Its speed, as a matter of fact, is extremely fast—over 186,000 miles per second.

## WAVELENGTH AND FREQUENCY

*Frequency* means the number of waves per second that leave the transmitting antenna or reach the receiving antenna. Radio waves move through space at a constant speed. The higher the frequency, the closer together successive waves will be. Saying it another way, the higher the frequency, the shorter the wavelength.

Oddly enough, radio waves moving through the atmosphere behave quite differently at different frequencies. This is due to the effect of the sun on the atmosphere, and it changes from day to night, as well as from year to year. Since amateurs are allowed to transmit a wide variety of frequencies, it is a good idea to become familiar with the basic characteristics of the different frequency bands.

## FREQUENCY CLASSIFICATION

The frequency of radio waves is expressed very broadly in

seven classifications: low (LF), medium (MF), high (HF), very high (VHF), ultra high (UHF), super high (SHF), and extremely high (EHF). Most of today's amateur operation takes place in the high, very high, and ultra high frequency ranges. Table 5-1 lists the classification of frequencies. Figure 5-1 shows the relationship between frequency and wavelength.

In spite of its extreme speed, the wavelength—the distance between succeeding waves—can be measured. Normally, this measurement is expressed in metric terms. Since higher frequency means shorter wavelength, an experienced person can determine the approximate frequency if he knows the approximate wavelength. The speed of radio waves was given earlier in miles per second. It comes out as a much more manageable number if we use metric terms—300,000,000 meters per second.

Frequency is normally expressed in millions of cycles per second. In recent times, hams adopted the word *hertz* to mean cycles per second. Thus, a million cycles per second is one *megahertz* (mega means million).

If the waves are moving out of the antenna at a million per second, and traveling at a rate of 300 million meters per second, it is easy to see that succeeding waves should be 300 meters apart. A frequency of one megahertz, then, has a wavelength of 300 meters.

The longest wavelength used by amateurs is 160 meters, somewhere in the neighborhood of 500 feet; and some amateurs operate with wavelengths less than a foot. Most of us, however, are somewhere between the two extremes.

**Table 5-1. Frequency Classification According to Wavelength.**

| Nomenclature | Abbreviation | Min. Value | Max. Value |
|---|---|---|---|
| Very low frequency | VLF | 10 kHz | 30 kHz |
| Low frequency | LF | 30 kHz | 300 kHz |
| Medium frequency | MF | 300 kHz | 3 MHz |
| High frequency | HF | 3 MHz | 30 MHz |
| Very high frequency | VHF | 30 MHz | 300 MHz |
| Ultrahigh frequency | UHF | 300 MHz | 3 GHz |
| Superhigh frequency | SHF | 3 GHz | 30 GHz |
| Extremely high frequency | EHF | 30 GHz | 300 GHz |

**LEGEND:** kHz = kilohertz (thousands of hertz)
MHz = megahertz (thousands of kilohertz)
GHz = gigahertz (thousands of megahertz)

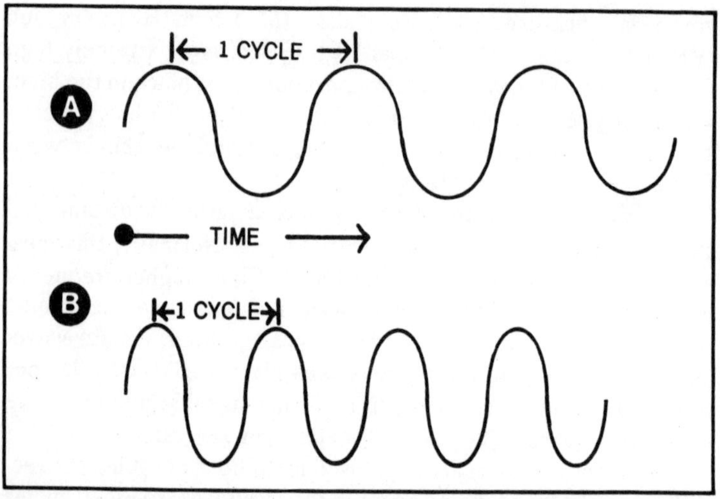


Fig. 5-1. Radio waves come in different sizes. Some have only a few peaks and valleys (cycles) per second, while others have many such cycles. The term for cycles per second is *Hertz*, abbreviated Hz, in honor of an early experimenter in radio wave phenomena. An amateur frequency of 3.7 MHz means that the wave has 3,700,000 cycles per second; amateur bands range from 1.8 MHz up to more than 1,215 MHz.

Those portions of the radio spectrum allocated to amateurs are called *bands*. Generally speaking, most of the amateur bands are harmonically related. That is, the frequencies are approximate multiples of one another. Table 5-2 shows the amateur bands and the operating privileges of each. Let's examine the ways the different bands behave.

**Table 5-2. Transmission Modes, and the Frequencies They Can Be Used On. (Continued Through Page 36.)**

| FREQUENCY MHz | BAND, meters | MODE AUTHORIZED |
|---|---|---|
| 1.8—2.0 | 160 | carrier-keyed telegraphy |
| | | amplitude-modulated telephony |
| 3.5—4.0 | 80 | carrier-keyed telegraphy |
| 3.5—3.775 | | freq-shift-keyed telegraphy |
| 3.775—4.0 | 75 | amplitude-modulated telephony |
| | | frequency-modulated telephony |
| 3.775—3.890 | | AM or FM television |

| | | |
|---|---|---|
| 7.0—7.3 | 40 | carrier-keyed telephony |
| 7.0—7.1 | | freq-shift-keyed telegraphy |
| 7.075—7.1 | | amplitude-modulated telephony |
| | | frequency-modulated telephony |
| 7.15—7.225 | | AM or FM television |
| 7.15—7.3 | | amplitude-modulated telephony |
| | | frequency-modulated telephony |
| 10.1 - 10.15 | 30 | carrier-keyed telegraphy<br>freq-shift keyed telegraphy |
| 14—14.35 | 20 | carrier-keyed telegraphy |
| 14—14.2 | | freq-shift-keyed telegraphy |
| 14.2—14.275 | | AM or FM television |
| 14.2—14.35 | | amplitude-modulated telephony |
| | | frequency-modulated telephony |
| 21.0—21.45 | 15 | carrier-keyed telegraphy |
| 21.0—21.25 | | freq-shift-keyed telegraphy |
| 21.25—21.35 | | AM or FM television |
| 21.25—21.45 | | amplitude-modulated telephony |
| | | frequency-modulated telephony |
| 24.89 - 24.99 | 12 | carrier-keyed telegraphy |
| 24.89 - 24.93 | | freq-shifted keyed telegraphy |
| 24.93 - 24.99 | | AM or FM television<br>amplitude-modulated telephony<br>frequency-modulated telephony |
| 28.0—29.7 | 10 | carrier-keyed telegraphy |
| 28.0—28.5 | | freq-shift-keyed telegraphy |
| 28.5—29.7 | | amplitude-modulated telephony |
| | | frequency-modulated telephony |
| | | AM or FM television |
| 50.0—54.0 | 6 | carrier-keyed telegraphy |

| FREQUENCY MHz | BAND, meters | MODE AUTHORIZED |
|---|---|---|
| 50.1—54.0 | | amplitude-modulated telephony |
| | | AM facsimile |
| | | frequency-modulated telephony |
| | | freq-shift-keyed telegraphy |
| | | modulated AM telegraphy |
| | | modulated FM telegraphy |
| | | AM or FM television |
| 51.0—54.0 | | unmodulated carrier |
| 144.0—148.0 | 2 | carrier-keyed telegraphy |
| 144.1—148.0 | | unmodulated carrier (AM or FM) |
| | | modulated AM telegraphy |
| | | amplitude-modulated telephony |
| | | AM facsimile |
| | | AM or FM television |
| | | freq-shift-keyed telegraphy |
| | | frequency-modulated telephony |
| 220—225 | 1¼ | All AM and FM forms of transmission |
| 420—450 | ¾ | |
| 1215—1300 | | |
| 2300—2450 | | All AM and FM forms of transmission, plus "pulse." |
| 3300—3500 | | |
| 5650—5925 | | |
| 10,000—10,500 | | All AM and FM forms of transmission |
| 21,000—22,000 | | All AM and FM forms of transmission plus "pulse." |
| ABOVE 40,000 | | |

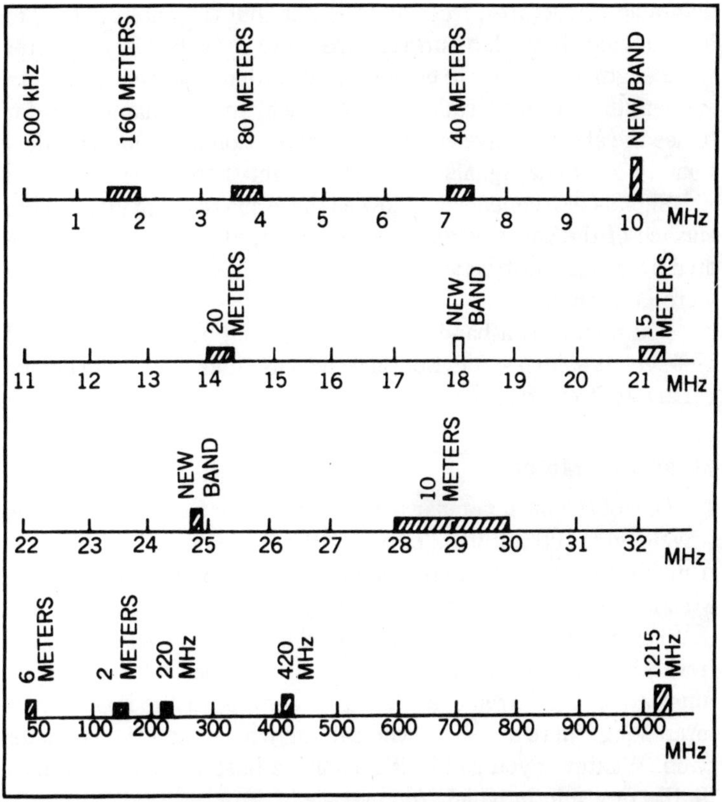


Fig. 5-2. The radio frequency spectrum covering the major amateur bands. Standard AM broadcast stations are in the region of 550 to 1.6 MHz; FM broadcast frequencies range from 88 to 108 MHz; and television stations use from 56 to 88, 174 to 218, and 470 to 890 MHz. Amateur bands are sandwiched in between many other services, including international shortwave, government, and private communications, and radar and position-locating signals. Note the new bands at 10, 18, and 24 MHz. These are a result of the World Administrative Radio Conference (WARC) held in Geneva, Switzerland in the fall of 1979.

## 160 Meters

This is the lowest frequency with the longest wavelength of all the amateur bands. The frequency range extends from 1800 kilohertz to 2000 kilohertz. For a long time, this band was shared by amateurs with LORAN radionavigation signals. Times have changed, and now the entire band is available to us. Not all amateur equipment covers this band, so operation here is comparatively limited.

Extremely long waves are easily diffracted by air, mountains, man-made structures, trees and the like, but they are too long to be reflected. Indeed, a surface capable of reflecting a 160-meter wavelength would have to be much greater than 160 meters in width and length. It is understandable, then, why many amateurs prefer to use the shorter wavelengths, as they rely on ionospheric reflections to get their signals where they want them.

Signals of frequencies as low as these snake along close to the surface of the earth, where they are bent, absorbed, and attenuated by surface obstructions not too many miles from the place they were generated.

160 meters is a band of relatively local communication, and quite vulnerable to static, But those who use it love it for the unique challenge it offers.

## 80 and 75 Meters

One of the most popular bands is twice the frequency (half the wavelength) of the 160 meter band. It covers a portion of the spectrum wide enough to run roughly from 85 to 75 meters (3.5 to 4.0 MHz). The lower segment, exclusively allocated to radio telegraph (CW) operation in this country, is called the *80 meter band*. It ranges from 3.5 to 3.75 MHz. The remainder, used primarily for voice communication, is referred to as the *75 meter band*. Just to make things interesting, there are some who call the whole thing the 80-meter band. Whatever you call it, it's always a bustling band, and may be the first one in which you operate.

At 75 meters, waves are still a bit too long for effective ionospheric reflection, but they are still short enough to resist attenuation for a longer period, and may typically be propagated over longer distances.

During the day, static and man-made noise are quite prevalent, especially during the summer. Color TV sets radiate a signal at approximately 3.575 kHz, which comes in loud and clear on nearby amateur receivers. Still, contacts up to a thousand miles or so are commonplace. At night the band opens up and stations several thousand miles away can be heard. The band is quite crowded during the night.

## 40 Meters

40 meters offers the best of two worlds. This band extends from 7.0 to 7.3 MHz, and the wavelength is short enough that signals can be reflected by the upper atmosphere (ionosphere), but they are also long enough to allow good groundwave characteristics.

Ionospheric reflection is good enough to carry signals halfway around the world under the right conditions, and, at sunrise and sunset, on the dark side, they come pouring in from all over the world.

The excellent propagation conditions on this band can be both a blessing and a curse. You see, 40 meters isn't reserved exclusively for amateurs in other countries as it is in the United States. Foreign propaganda broadcasts sometimes make shambles out of operation, but this is nonetheless a very popular (long-distance) DX band.

### 30 Meters

The 30-meter band (10.1 to 10.15 MHz) is a relatively new one, having come into existence just a few years ago. It is presently authorized only for code or teletype operation with a power limit of 250 watts. There are still a few fixed government stations operating in this band, and they have a priority claim to the frequencies they use. However, a "forbidden zone" from 10.109 to 10.115 MHz has recently been eliminated.

This band lies between two of the hottest DX bands, and shows a lot of promise for the future.

### 20 and 15 Meters

These amateur bands offer some spectacular opportunities to show what radio can do, but successful operation depends on where the sun is, and how active solar storms are. Sometimes you may sit at your rig for hours and hear nothing. At other times, conditions are so good you can't even find a hint of a clear spot in which to operate.

Amateurs who frequent the 20-meter band (14.0 to 14.35 MHz) and the 15-meter band (21.0 to 21.45 MHz) quickly learn to recognize the signs of optimum propagation characteristics. They learn that certain years are better than others for long-distance communications, as the high points of activity correspond to the 11-year sunspot cycle. They also learn that the time of day has a lot to do with successful operation, and that a dead band at 2:00 PM is no indication that the band won't be alive soon after.

### 17 and 12 Meters

These are two of the new bands that were approved for amateurs at the World Administrative Radio Conference in 1979. As of this writing, one is still not yet available, but the machinery is in motion. The 17-meter band, 18.068 to 18.168 MHz will not become available for quite a while, as other, more vital services are using it. However, as this text was going to the printer, the 12-meter band (24.89 to 24.99 MHz) became available, with General class and higher being allowed CW operation over the entire band, and single-sideband voice communication allowed above 24.930 MHz. Teletype is allowed up to 24.930 MHz. This band will have some excellent DX at the peaks of propagation activity, but at other times will be more local.

### 10 Meters

The 10-meter band is big as far as spectrum space is concerned. It is very much like the CB frequencies in its behavior, being very dependent on sunspot activity for DX operation. Hams flock to this band at the peaks of activity and manage phenomenal contacts with unbelievably low power output levels. Then, as the solar cycle progresses and conditions become less favorable, the band becomes almost deserted, monitored by the faithful few who patiently wait for those freakish conditions that result in rare DX contacts.

This band is the upper limit of that range of frequencies called *high frequency*. Above 30 MHz, it's called *VHF*, or *very high frequency*.

### 6 Meters

The lowest frequency VHF amateur band is the 6-meter band. It covers a frequency range from 50 to 54 MHz. Normally, this band is used for local work, as the waves tend to propagate chiefly along line of sight. There are times, however, when ionospheric reflection occurs, treating you to some rare DX. These occurrences are coincident with sunspot activity, and an operator may work this band for years without getting more than 50 or 100 miles.

6 meters is immediately adjacent to the television channel 2, and activity on this band is understandably light in cities where that channel is in use. Those who do operate this band in cities that use Channel 2 need much savoir faire in handling neighborhood TV interference complaints. Also, this band is right in the middle of a natural spectrum of electrical noise, and is very susceptible to ignition, arcing at power poles, generator whine, and other man-made pains in the neck.

### 2 Meters

Presently the most popular VHF band, 2 meters covers the

range of frequencies from 144 to 148 MHz. Propagation conditions here are similar to those on 6 meters, but DX is even more rare. However, this rarity, coupled with the relatively small size of antennas and the relative freedom from electrical noise makes it an extremely popular band for local work. Most of the voice transmission here is frequency modulated.

By mutual agreement, hams have divided portions of the band into channels, some being designated for specific kinds of operation. This is a voluntary action, not a regulation. Hand-held walkie talkies are extensively used in this band, assisted by automatic repeaters that retransmit the signals for far more efficient local operation.

### Beyond 2 Meters

Frequencies above 2 meters include two large bands, one from 220 to 225 MHz, and one from 420 to 440 MHz. There are many bands at higher frequencies, but the variety of conditions exists mostly at lower frequencies. These bands are chiefly line-of-sight bands. Once communication is established, however, reliability is very high. If you contact an amateur one day, you can be sure of reaching him the next.

## FINDING THE ACTION

Amateurs do much more than general rag-chewing, and these activities can be a lot of fun if you know where to look for them. Table 5-2 lists the types of signals that can be used on the different bands. For the beginner who wants a neat, well-keyed signal to practice code reception, the ARRL headquarters station, W1AW, transmits code practice lessons on almost all bands. The most popular spot is 3.580 kHz, close to that distinctive, rippling signal radiated by your color TV.

Between 3.6 and 3.7 MHz you hear many teletype stations. Also, if you can copy fast code, you can find numerous *traffic nets* exchanging third-party messages. This is one nice thing about amateur radio; you can send a free radiogram anywhere in the Western hemisphere.

Next, of course, is the 80-meter Novice band, the only one that has remained unchanged since the Novice license was instituted. There is nothing to prevent a ham with a higher grade from operating within the Novice band. Most novices appreciate a contact with an experienced ham who keeps his speed down to that at which the novice sends, and is a bit understanding, remembering that he too was once a beginner.

Just above the Novice portion of 80, you find the favorite hangout of our Canadian neighbors. Tuning higher, you hear many nets doing everything from casual rag-chewing to playing chess, studying the Bible, exchanging weather information, or passing messages.

40 meters is a favorite international band. You find a few teletype stations in the CW portion, and a few traffic nets. In the late afternoon and early evening, you can often hear foreign missionaries talking with home.

15 and 20 meters are the bands for the DX-seekers. There are a number of missionary nets here as well. The phone (voice) portion of the band is a favorite spot for slow-scan TV. You may hear hams on opposite sides of the country exchanging pictures! The pictures are still photos, rather than the moving pictures of conventional television, and they aren't as sharp in detail. However, they use no more spectrum space than a single-sideband voice transmission. Conventional, fast-scan television uses 2,000 times as much spectrum space.

On ten meters we find some even more exotic activities. Some of the OSCAR satellites, orbiting the earth, receive 2 meter amateur signals and reradiate them on 10 meters. With the aid of these satellites, amateurs communicate around the world using only a few watts of power, and frequencies that normally wouldn't get over the horizon.

On 2 meters and above, you find more satellite communication, and occasional moonbounce activity—exactly what it implies. Amateur signals are bounced off the moon, and can result in a two-way communication halfway around the world.

Frequency modulation is the most popular voice communication mode on 2 meters. Also, while there are occasional repeaters on 10 and 6 meters, most of them are on 2 meters.

The higher bands have similar activities to the 2 meter band, except most operation above the 430 MHz band is largely on the experimental basis.

If you tune just above the upper limit of the HF amateur bands, you will hear MARS stations operating. Although MARS stations are run by hams, do not try to transmit on those bands unless and until you become a MARS member. Once you acquire MARS membership, you will be welcomed on these frequencies. Before that, you'll be told to get back in the ham band.

# Chapter 6
# Etiquette and Operating Practices

AN AMATEUR OPERATOR HAS A CERTAIN RESPONSIBILITY to his fellow hams when he transmits a signal from his station. Amateur radio attracts the kind of people it does primarily because hams generally behave with a reasonable degree of responsibility on the air. If you are to be successful and accepted on the air, there are certain things you simply do not do. Sure, you will find a certain number of turkeys among your fellow hams. That kind exists in any society. However, their misconduct does not justify yours.

Some of the rules of responsible behavior are legislated by the FCC rules and regulations; the rest come from a voluntary code of ethics mutually agreed upon by the amateur radio community. If you are new to the hobby, I suggest two things: First, read Part 97, subparts D and E of the FCC rules and regulations. This will help keep you out of trouble with the FCC. Second, listen to other operators who use the frequency bands you are most interested in before you put your own rig on the air.

## STATION IDENTIFICATION

Just like any commercial radio station, your amateur station must be identified at regular intervals when you are using it. And you must use your assigned call letters, not a "handle." Generally speaking, handles are taboo with amateur radio. I say this even though I sometimes use one myself. I come on a net with "This is the Fez—W2FEZ." I get away with it because I immediately give my call letters. If I should forget, I am quickly reminded.

You must identify your station when you first turn on your station for an operating session. It isn't hard to do. Simply say something like "This is W2FEZ. Is this frequency in use?" (The query is a rule of courtesy which we'll discuss later.) If you were operating in CW, you would send DE W2FEZ. DE is a world-recognized means of saying, "My call letters are . . . "

When you contact another ham, you say, "K6MVH, this is W2FEZ." In CW you would say K6MVH DE W2FEZ. You should identify yourself in this manner at the beginning and end of each transmission, and if your transmission is a long one, you should identify every 10 minutes during the transmission. The exception is if you're making many brief transmissions; then you just identify every 10 minutes. Transmissions like "K6MVH, this is W2FEZ. What did you say? K6MVH this is W2FEZ." are both unnecessary and ridiculous. The bottom line is that both you and the other operator must identify at the start and finish of such a series of exchanges, and every 10 minutes during your conversation.

If you are operating somebody else's station, give both call signs in this manner: "This is WA2JHD with W2FEZ at the mike." In code, you would send DE WA2JHD/W2FEZ.

If you take your station to another location, or if you operate a station in your car, you are obliged to give your location along with your call letters, and state either portable (fixed location away from home), or mobile.

## GENERAL CALLS

Sometimes you'll turn on your station and listen for a while. You may hear several stations communicating with one another. Then, if you listen long enough, you'll hear somebody calling CQ. CQ is a general call that means the transmitting station is looking for someone to talk to. If you don't hear it, and you want to talk to someone, you might make such a call itself.

If you listen to other amateurs before you start operating your own station, you will notice that the call, together with the accompanying information, tends to have a definite form. Ideally, one should repeat the CQ three times, followed by the calling station's identification, and repeat this whole formula not more than two or three times before listening. Sometimes a beginner, discouraged by poor results, will give out a long-winded CQ, little realizing that he is driving away the best prospective contacts.

My call is K6MVH. If I were making a general call, this is what I would say:

"CQ, CQ, CQ. K6MVH calling CQ. CQ, CQ, CQ. This is K6MVH. Kilowatt 6 Mike Victor Hotel calling CQ and listening."

Then I would listen carefully on my own frequency for a short period, then tune slightly above and below my own frequency to see if another amateur was answering my call.

Notice the use of phonetic expressions to clarify my call letters. This is the way it's done with voice transmissions. Some amateurs make up a slogan to go with their call letters, instead of using the standard phonetic alphabet. For example, WA2JHD calls himself WA2 Just Had Dinner, and you can imagine the phonetics used by K2DWI!

If I were calling in CW, it would go like this:

CQ CQ CQ DE K6MVH K6MVH CQ CQ CQ DE K6MVH K6MVH AR K

On two meters FM, the technique is different. I have never heard anybody call CQ through a repeater. As I mentioned in an earlier chapter, hams have channelized two meters by mutual agreement. There they say something like this:

"This is W2FEZ listening on channel." or "This is W2FEZ listening on the repeater."

If anybody is listening to the repeater, he or she will hear that and respond. If nobody answers, just listen to the repeater for a while, and you'll hear somebody else call in a similar manner.

## CALLING A SPECIFIC STATION

Suppose you hear a station calling CQ, and want to talk with him, or suppose you have arranged to meet another ham on the air at a specific time. Now, the fellow who called CQ will be looking for somebody to call him; the one with whom you have a schedule knows you're going to call him. Therefore, there is no need to make a long-winded call. As with calling CQ, keep it short. Give the other person a brief call, and follow it with a brief standby period. If he doesn't come back to you, try again.

If I were making a specific call, I'd do it like this:

"WA2NSD, WA2NSD, WA2NSD. This is K6MVH calling WA2NSD. Kilowatt 6 Mike Victor Hotel calling WA2 Never Say Die. K6MVH calling WA2NSD and listening on frequency."

The format isn't necessarily sacred, but the idea is to keep it short. After calling, I listen for a minute or so on frequency, and check about a kHz or so either side of my frequency, just in case there is a slight difference between transmit and receive frequencies of the two stations. If band conditions are bad, or there are a lot of potential interference, I call perhaps half again as long.

## LOG KEEPING

Amateurs were originally required to keep a log of their activity. Not so any more—this requirement was done away with several years ago. Nonetheless, many hams still do. A meticulous log is one of the marks of a good operator. Moreover, the log is good to examine in later years, remembering some whom you haven't heard from in a long time.

Figure 6-1 shows a typical page from an amateur's log.

## THE NO-NO'S OF HAMMING

As with anything, some things are forbidden, and others may offend other hams. Subpart E of the FCC Rules & Regulations has the specific details on what you can and cannot do. You would do

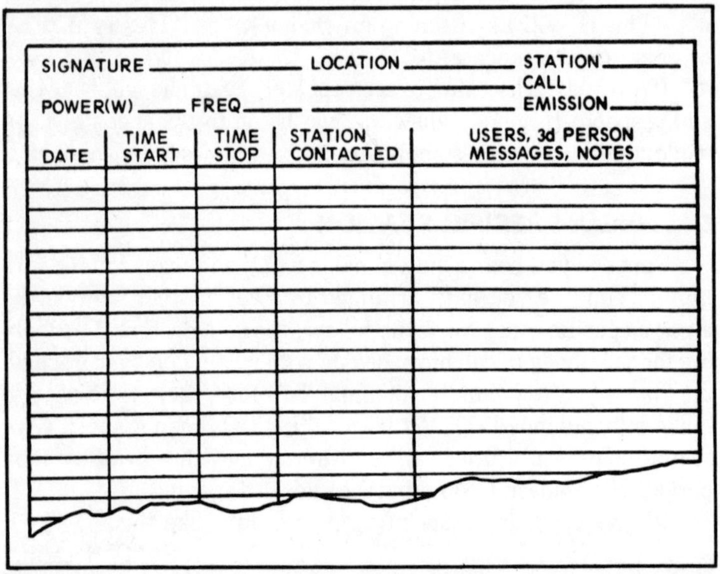


Fig. 6-1. This typical logsheet allows sufficient space for accumulating notes about amateurs contacted, and other information to jog your memory at some later date.

well to read it, especially since you must be familiar with the rules in order to pass your license exam. Briefly, however, here are some of the things that are taboo.

### Charging Money For Your Services

The services that amateur radio extends to others have always helped give our hobby high public esteem. Our very name implies that we are into this thing for the love of it. It is in keeping with the rules, as well as in our better interest, to keep it that way. The government thinks so, too. You can't use your station for a profit; you can't take money, and you can't accept gratuities. The rules are very explicit on this: "An amateur station must not be used to transmit messages for hire, nor for communication for material compensation direct or indirect, paid or promised." You can't say it much more clearly than that.

Ham radio is a hobby. There are other services in which stations are licensed for the business or communicating. Let the business people handle the business. Let's stick to the fun of ham radio.

### Broadcasting

We are licensed for two-way communication. You cannot use an amateur station for broadcasting to the general public. However, there are certain kinds of one-way transmissions that are permitted. Part 97.91 of the Rules says, ". . . the following kinds of one-way communications, addressed to amateur stations, are authorized and will not be construed as broadcasting: (a) Emergency communications, including bona-fide emergency drill practice transmissions; (b) Information bulletins consisting solely of subject matter having direct interest to the amateur radio service as such; (c) Round-table discussions or net-type operations where more than two amateur stations are in communication, each station taking a turn at transmitting to the other station(s) of the group; and (d) Code practice transmissions intended for persons learning or improving proficiency in the international Morse code."

### Music

Even if you aren't broadcasting to the general public, and even if you feel your singing or playing is so bad it doesn't qualify as music, you still can't do it. You can't subject others to your dabbling. The FCC prohibits amateurs from singing, whistling, playing instruments, or transmitting recorded music. You can use a single-tone whistle to enable you to tune your rig, but that's all. Don't try singing "Happy Birthday" to your mother.

### Obscenity

This is a sticky subject. Even if we all agreed as to what is obscene and what isn't, promoters of obscenity would try to redefine the word to their own ends. The Rule forbids obscenity, indecency, and profanity, without defining the words. Obscenity lovers would insist that each and every word or phrase that is obscene be spelled out in the rules, but the rules do not do that. Consequently, the courts are often pressed for judgments as to where decency ends and profanity begins. That which is holy to me might be contemptuous to you; that which is indecent to me might merely be "colorful" to you. Where do you draw the line?

Personally, I prefer not to come near the line. As a general rule, you might follow the example of the average, middle-of-the-road broadcasting station. I don't know of any hams that got into trouble with an occasional "hell" or "damn." Still, if you used anything beyond that to excess, you would find yourself without anybody to communicate with. Whatever may be your belief as to word usage, consideration and respect for the feelings of others never hurt anybody. We hams aren't prudes, but we do respect others. It's nice to be nice.

### False Signals

Absolutely forbidden—even on the first day of April! When you identify your station, you must use the right call letters. You break the rules if you try to fool your friends into thinking you're somebody else, and you're way out of line to transmit a phoney distress call. They take that one very seriously, and it's spelled out quite clearly in the rules.

### Interference

You're bound to run into *some* interference. In today's crowded ham bands, it's a way of life. Just don't do it on purpose. Also, try to avoid accidental interference. Ham band courtesy insists that you listen before you transmit, and never operate on a frequency where others are using it first, unless, of course, you want to join them. Even then, wait for a pause and call in with discretion. Amateur frequencies are available on a first-come, first-served basis.

The fact that a net agrees to meet regularly on a frequency doesn't give them claim to that frequency, even if they have been using it for years. On the other hand, if you know the net usually uses that frequency, you do rather poorly to fire up on the frequency a minute or so before you know they're about to meet. It's all a matter of give and take, and remembering that nobody owns a frequency.

### Confidentiality

Radio communications are private. They always have been. It is very wrong to divulge information you get off the air, except to the person for which it is intended. This applies both to message traffic and to the contents of casual conversation. Passing on intimate bits of "dirt" you overhear on the air can get you into big trouble.

## MESSAGE HANDLING

It's almost certain that, sooner or later, you'll be involved in passing a message on for somebody else. Amateurs have always done that. In fact, the very name of the biggest amateur organization of all implies the relaying of messages to be its major function.

There is a loosely-organized but extremely efficient system for passing messages by amateur radio, and every day they pass thousands of pieces of traffic all about the country. When you consider that it's an all-volunteer system, and that it's 80% or better efficient at getting messages to their final destinations, it becomes rather impressive.

The success of this activity is due to the fact that they have and stick to a standard procedure and message format. Local nets pass traffic within their own areas of coverage, and then check into regional nets where messages are passed from one net to another. The regional nets communicate, at a higher level, with other nets within a time zone. Finally, there are those who pass traffic between the time zones. A message going across the country could originate in a local net, be passed to a regional net, then to a section net. From there, it would go to a section net in the proper time zone, to a region net near its destination, and from there to a local net. The local net would pass it to the member nearest its destination who would deliver it. All this for free.

The nets are directed by a *net control* station, who acts as moderator. Except under very urgent circumstances, it is considered very rude to break into a net without invitation from Net Control. Standing by for additional check-ins is part of standard net procedure. In code, the invitation is "QNI." Stations checking in will either say "QRV," meaning they have no messages to send, or "QTC (so many)." In voice, they might either say "No traffic," or "One routine for (whatever it's going)."

If a station checks in with traffic, the Net Control station directs that station to call either the station near the destination, or the station that represents that net to a higher level net.

Here is an example of a typical message format:

NO. 21R W2FEZ AVON NY 2100 DEC 21 1985
MOSES VAN CAMPEN
20 WILLETS AVE
BELMONT NY
FONE 337 7167

HAVE A MERRY CHRISTMAS X SEE YOU NEXT WEEK
JOHN BUTLER

All those letters and numbers at the start of the message have a purpose. "NO. 21R" identifies the message as number 21 from the station that originated it. The "R" places it at routine priority. The call letters that follow are those of the originating station. The "9" is a count of words in the message.

Note that there is no punctuation. The letter X takes the place of a period, and is counted as a word. After the count of words come the place, time, and date of origin.

Finally, a telephone number is included in the address, which makes the message easier to deliver.

One last point—if you are sending the message by voice rather than code, you should write down each word as you say it. Then you know you aren't talking faster than the other operator can write.

## THE PHONETIC ALPHABET

The phonetic alphabet is a standard set of words used to clarify messages being sent over a voice modulated station. Each word stands for a letter of the alphabet. For example, the name SMYTHE might be sent this way: "Smythe, I spell, Sierra Mike Yankee Tango Hotel Echo, Smythe."

This is the widely-used, internationally-accepted phonetic alphabet:

| | | | |
|---|---|---|---|
| ALPHA | GOLF | NOVEMBER | UNIFORM |
| BRAVO | HOTEL | OSCAR | VICTOR |
| CHARLIE | INDIA | PAPA | WHISKEY |
| DELTA | JULIETTE | QUEBEC | XRAY |
| ECHO | KILOWATT | ROMEO | YANKEE |
| FOXTROT | LIMA | SIERRA | ZULU |
| | MIKE | TANGO | |

These words are more or less standard, but they are by no means sacred. Often amateurs will adopt a catchy phrase to go with their call letters, as we pointed out earlier in this chapter. The important thing is that the meaning is received correctly at the other end.

# Chapter 7
# Neighborhood Interference

THERE IS PROBABLY NOT AN AMATEUR OPERATOR ANYwhere in any residential or metropolitan area who has not experienced interference between his station and his neighbor's TV, radio, or stereo. The degree of seriousness of such a problem depends largely on the amateur and the way he conducts himself when the neighbor first confronts him with the problem. This often has a greater bearing on the eventual solution of the problem than determining who is at fault. Courtesy and consideration of the other's feelings are always of prime importance.

## TVI AND BROADCAST INTERFERENCE

TVI is and always has been a major cause of concern to every amateur. When television broadcasting saw its first widespread public use in the late 1940s and early 1950s, relatively little was known about the many factors that cause this problem. Neither the entertainment electronics industry nor amateur radio was prepared for the head-on collision between broadly-tuned TV sets with microvolt sensitivity and the relatively high harmonic output of the Class C amplifier, which was then the standard output stage of every radio transmitter in existence. Interference from amateur stations was so widespread that the entire ham radio fraternity was put in a bad light.

However, amateurs quickly discovered the sources of the problem and learned to eliminate them, so far as transmitting equipment was concerned. This was done so effectively that today 95% of all TVI complaints received by the FCC are traced to design deficiencies in the TV receivers.

Through the 1950s, many TV manufacturers were concerned only with producing the sets as cheaply as possible, with no concern for designing them to reject unwanted signals. This resulted in serious interference problems that were not the fault of the amateur stations. TV owners who had paid hundreds of dollars for their sets didn't want to hear it when they were told that the trouble was inherent in their equipment. There was a lot of ill feeling, and even some civil lawsuits. Then the FCC cracked down on the TV manufacturers and required them to either design the sets right in the first place, or to make additional filtering available to the customer on request.

Today the situation is much improved, but interference can still happen. An amateur transmitter of modern design usually will not cause interference unless it or its antenna is improperly tuned. As to the TV manufacturers, we can understand their unwillingness to attack the problem when we realize what a bucket of worms it can be. Consider these facts:

One TV channel occupies six times as much space in the radio spectrum as the entire AM broadcast band; 2000 times as much as an amateur SSB voice communication—more than all the amateur bands below 6 meters combined.

Because of this wide signal, the front-end tuning of a TV set has to be broader than a barn door. The 6-meter amateur band is within the passband of a TV set tuned to Channel 2.

TV signals are of such a nature that they have to be 100 times the strength needed for an AM broadcast signal to produce satisfactory reception.

We can better understand the mechanism of TVI if we examine the possible sources of the problem. Some interference happens simply because the amateur is operating on a frequency that is too close to that of a television channel. For example, the top end of the 6 meter amateur band is the bottom end of Channel 2. Unless the front end of the television receiver is particularly selective, a healthy signal from a nearby amateur operating on 6 meters will brute-force its way into the TV receiver, causing havoc with the picture.

Some TV sets of the mid 50's had intermediate frequencies that were either within or harmonically related to certain amateur bands.

This was an open invitation to interference. Also, some harmonics (multiples) of amateur frequencies just happen to fall within television channels.

The spectrum chart (Fig. 7-1) shows us that many amateur bands are uncomfortably close to radio and television broadcast frequencies. For instance, the 160 meter band lies just above the AM broadcast band; 2 meters is between the FM broadcast band and Channel 7, and the 420 - 450 MHz amateur band isn't too far from the bottom of the UHF television band.

In addition to these possible sources of adjacent-channel interference, we can be plagued by harmonic interference. Harmonics of the 80- and 40-meter bands fall in all of the VHF TV channels. Harmonics of certain parts of the 80-meter band fall in the 40-, 20-, and ten-meter bands. Harmonics of 40 meters fall in the 20 and 10 meter bands. Harmonics of the 20-meter band fall in 10 meters and in TV Channel 2. (Channel 2 is 6 MHz in bandwidth.) Harmonics of 10 meters fall in TV Channel 2 and in the 2-meter amateur band. Harmonics of 6 meters fall in the FM broadcast band and in the middle of TV Channel 11.

When you consider all this, coupled with the possibilities of amateur harmonics falling in the intermediate frequencies of commercially produced equipment, you must agree that the amateur who completely escapes interference problems of any kind is nothing but lucky.

Although one occasionally runs into a particularly nasty TVI problem involving a very old set, fringe-area reception, or both, the overall interference problem is far from hopeless. The first step is to make every effort to stay on speaking terms with the neighbor who complains. This can be a bit tricky with an individual who understands little or nothing about radio communication, and is skeptical at best about your abilities.

If you can at least convince your neighbor that you, too, are concerned and anxious to solve the problem, that's half the battle. Next, you need to determine the cause of the interference. In some instances, it can be cured by more careful tuning of your equipment, a better antenna arrangement, better grounding, or adding a low-pass filter. However, there are many instances in which the problem must be tackled at the TV receiver.

If that is the case, your role is one of difficult diplomacy. *Whatever you do, never touch a tool to the TV!* I don't know how many amateurs have fallen into the trap of making "modifications" to their neighbor's TV sets. The invariable result is either that the

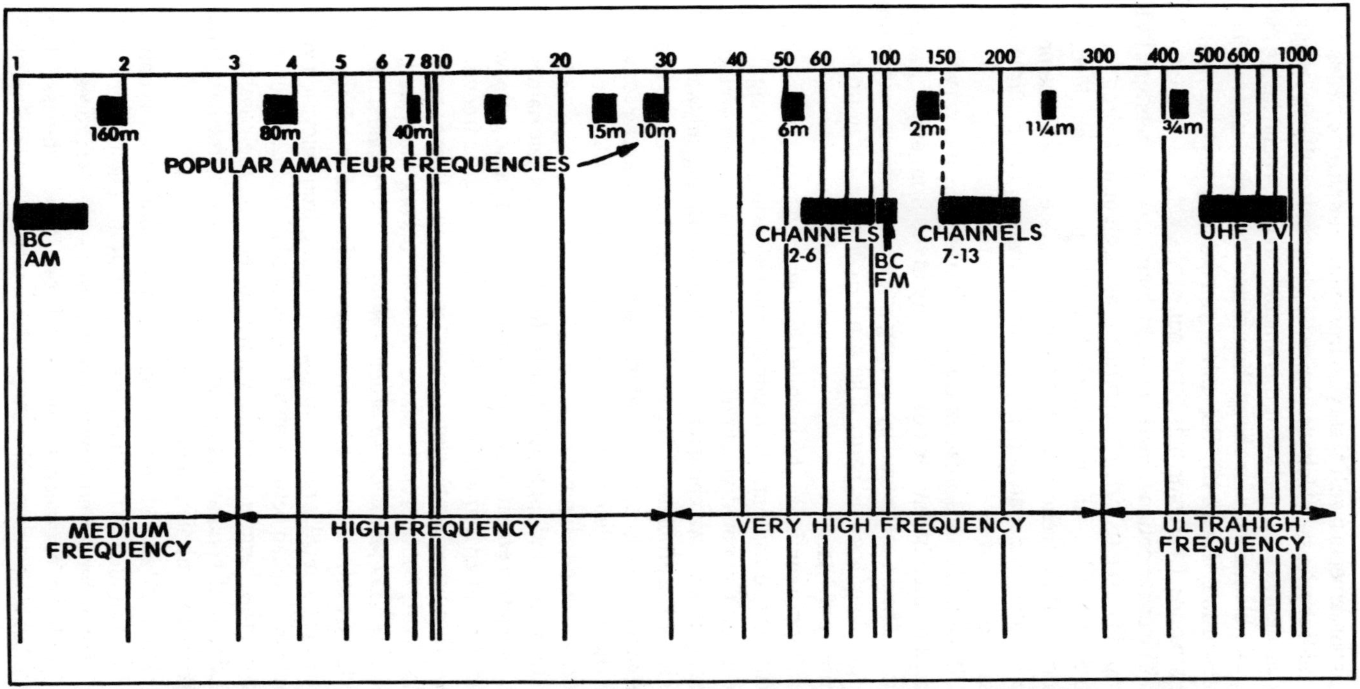


Fig. 7-1. Chart showing the location of TV channels and amateur bands in the spectrum from 1 to 1000 MHz.

neighbor expects free TV service from you ever after or, when the set malfunctions—which, of course, it ultimately will—the neighbor suddenly remembers that this amateur fooled around inside the set, and blames you for the malfunction. "It worked all right before you fooled with it." So, unless you want to pay your neighbor's future TV repair bills, stay out of the set.

The conscientious amateur should, long before he gets any TVI complaints, see to it that his station is as clean of undesired radiation as it can be. His equipment should be properly shielded, as commercial rigs generally are. The antenna system should be properly installed and matched, and he should have a good-quality low-pass filter.

If the problem is adjacent channel interference, it is very likely caused by an inherent lack of selectivity in the front end of the receiver. The amateur can, by making a few concessions, minimize the problem; the neighbor, if he can be convinced, can probably eliminate it.

You can minimize the problem by lowering your power, or by orienting your antenna to minimize radiation in your neighbor's direction. You could voluntarily stay off the air when you know your neighbor is watching a favorite program. (Do that anyway until you solve the problem—it's good PR.) Better than all these, if your neighbor is willing to come half way to meet you, you can convince him that the problem is in his set and that, if he installs a high-pass filter, it can be eliminated.

Low-pass and high-pass filters do exactly as their names imply. One lets frequencies below a predetermined value go by while blocking higher frequencies. The other allows higher frequencies to get through while blocking lower frequencies. There are plenty of low-cost, commercially made filters that work quite well, but remember: *Let your neighbor install it himself. Don't install it for him.*

If the problem proves traceable to interference on a frequency that is harmonically related to your operating frequency, it's your problem. You had better stay off the air until it's fixed. A local ham club or other hams in the area will very often offer their assistance in fixing it. If you haven't already done so, install a low-pass filter in your transmission line.

The diagrams shown in Figs. 7-2 and 7-3 are shown only to give you an idea as to what's inside the filters. The proper selection of inductance and capacitance values is a precise process. Unless you're an engineer, it's best to buy one. The high-pass filters that most hobby electronics shops sell for two or three dollars work quite well. (Don't buy a filter for your neighbor—let him buy it.) However, if the set is new and your neighbor is the first owner, he can generally get a pretty good filter either free or for a very low price by writing to the manufacturer of the set. He should specify the model and serial number of the set, and when he bought it. When he gets the filter, he should install it according to the instructions that come with it. The filter won't work unless it's properly installed.

For a low-pass filter for your own installation, there are several good makes, and Drake produces what many feel is the top of the line. Don't try to use a filter designed for CB use unless its manufacturer guarantees it can handle amateur power levels.

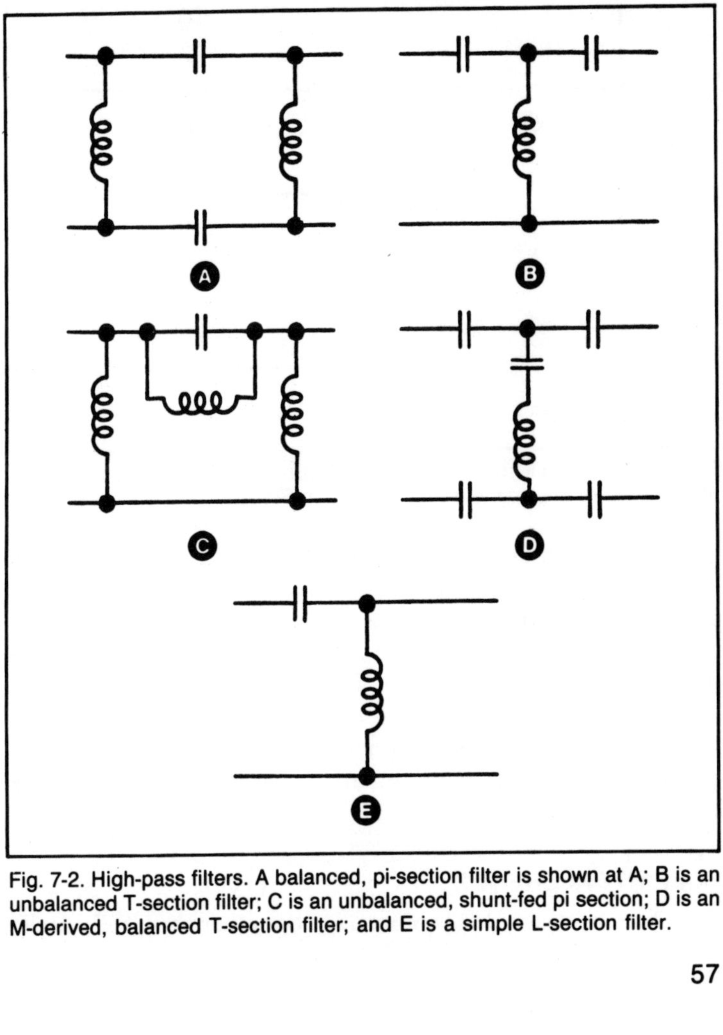


Fig. 7-2. High-pass filters. A balanced, pi-section filter is shown at A; B is an unbalanced T-section filter; C is an unbalanced, shunt-fed pi section; D is an M-derived, balanced T-section filter; and E is a simple L-section filter.

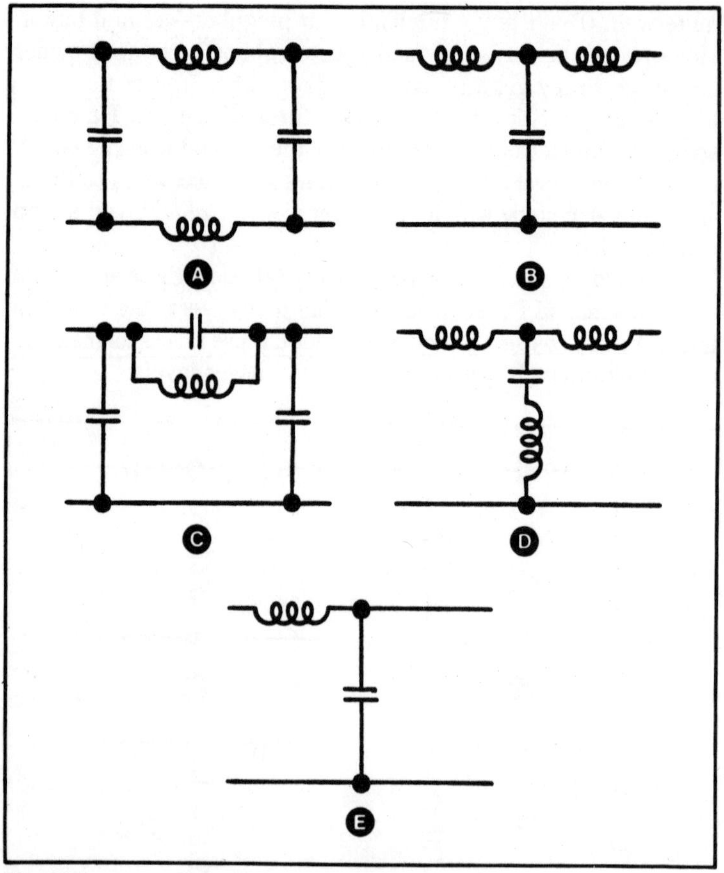


Fig. 7-3. Low-pass filters for transmitters. These types correspond to the high-pass filters of the same letter designation in Fig. 7-2.

Be sure your antenna is properly matched to the transmission line, and that the line impedance matches the design impedance of the filter. See that your equipment is well grounded. If you can transmit without interfering with your own TV, you can be reasonably sure your transmitter is clean. If the interference occurs on some sets but not on others, that is a good indication that the receiver is at fault. It is a good squelch for a blowhard to invite him over to your house to watch his program on your set, since his obviously isn't working right.

## HI-FI INTERFERENCE

There are many kinds of interference other than TVI. Audio

rectification in a stereo or a PA system is quite common, and is virtually always the fault of the audio equipment. The symptom here is that your transmission is heard on your neighbor's system. Complaints of peculiar operation of equipment that does not actually result in his hearing your signal, however distorted, are to be looked upon with suspicion. In severe cases, your signal will come pounding through and not be affected by the volume control.

An experience of mine gives a good example of this. I sat in a Salvation Army service in Batavia NY, one day, when a CB passing by almost broke up the meeting. The captain had just uttered the words, "Let us pray . . ." when a voice came clearly out of the PA system, "Breaker one nine." The next day I donated a capacitor to the amplifier to put an end to the problem.

Whenever your signal gets into the amplifier stages (not the radio stages) of your neighbor's audio equipment, bypass capacitors must be installed. Here again, the burden of getting the job done lies with the owner of the audio equipment, not with the amateur. His equipment is not supposed to be sensitive to signals in the radio frequency region—at least not when he's listening to records or tapes. Capacitors placed at strategic positions in the circuit will short the radio signal to ground while allowing the audio signals to pass.

## "MIXING" INTERFERENCE

Interference caused by the mixing of several frequencies is perhaps the hardest kind to pin down and eliminate. There are an unlimited number of devices that can combine two radio signals and produce a third as a result. This is a common phenomenon, and is utilized in practically every piece of radio equipment in use. It is evident when the interference comes and goes at irregular intervals, or your signal appears to be causing interference on a frequency that is neither harmonically related nor in a situation to cause adjacent channel interference.

When two dissimilar signals are fed into the same detector under the right conditions, a mixing action occurs. This causes a number of new signals to be produced, which are mathematically related to the combination of the original two. For example, if a 3-MHz signal mixes with a 144-MHz signal, signals will be produced at 141 (144 − 3) and at 147 (144 + 3) MHz. If more than two signals are involved, many mix products can be produced. The fastest cure for a mix is a quick change of frequency. The most effective cure is to tune out all but one of the offenders. A small change in frequency can move the mixing products out of harm's way; if one can be filtered out, better still.

## AM BROADCAST INTERFERENCE

This is one situation that is almost never the fault of the amateur transmitter. While a transmitter can malfunction so as to produce a spurious signal (spur), it very rarely happens. More often than not, there is another explanation. It was not too long ago that a friend was telling me his woes. "I've got a spur (spurious emission) right in the broadcast band," he moaned, "so I can't stay on much longer." We discussed such things as traps in his antenna system, etc., but it wasn't until a month or so later that the truth hit me.

"Joe," I said when we next talked, "Exactly what is the frequency of your spur?"

"1274," he replied (meaning 1.274 MHz).

I had carefully rehearsed this. "Now you just get out your calculator and tell me the frequency of the lady's local oscillator when her radio's tuned to that frequency," I continued.

"1729."

"Good. Now what's its second harmonic?"

"3458."

"Fine. Now what's the next higher frequency that that second harmonic can mix with to produce the 455 kHz intermediate frequency?"

"3913. Hey, wait a minute!"

Funny thing about that. 3913 just happened to be the frequency of the net on which Joe and I met most often. The interference was entirely the fault of the cheap design of the sweet little old lady's receiver. Joe wasn't obligated to do a thing. However, he did voluntarily stay off the air when his neighbor was listening to her one favorite program daily. After all, it's nice to be nice.

Another evidence of fault on the part of the receiver showed up when a neighbor accused me of wiping out police communications. It seemed I came in loud and clear on his monitor. I called the local police, and they listened while I transmitted. No interference. It was clearly a problem in the guy's monitor. Had it been my transmitter, I would have come in on all police receivers, not just the one. My neighbor probably had the cheapest brand money could buy.

Unless your signals interfere with *all* the receivers, it's probably not your fault. If your signal wipes out only the cheap ones, don't worry about it. Be considerate nonetheless. You have to live with your neighbors, no matter where the fault lies.

## ON THE OTHER HAND

You may find that, as excited as people get when your transmitter interferes with their equipment, they remain cool and collected when you tell them their TVs are coming in on your ham radio. They could care less, but it does happen. A color TV can radiate a mean signal close to 3.58 MHz that can render that frequency all but useless. This can bring you to a head-on confrontation. The service department of the TV and appliance store will claim nothing can be done about it, and it's not worth the bother of complaining to the FCC.

A computer can wipe out the low channels on nearby TV sets, or produce a whole spectrum of noise in a ham receiver, especially if the computer is connected to peripherals. Other than shielding, there is little that can be done.

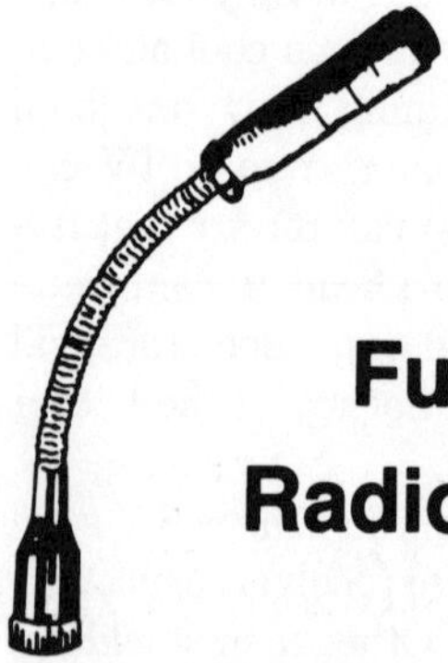

# Chapter 8 Fundamentals of Radio Communication

OF ALL THE BENEFICIAL USES OF ELECTRIC PHENOMENA, few have had a greater impact on our society in general than radio communication. By it, we learn of important events, no matter where in the world they may occur, in a matter of minutes; we discover the ideas and customs of our fellow man in other cultures; and we can, in times of great need, summon the assistance of our near neighbors.

In civilized countries, radio communication affects the lives of practically every household. In the lesser developed nations, it still reaches far into the wilderness to encourage learning and elevate the general level of all peoples.

Nobody knows for sure exactly what radio waves are, although there are some very sound theories. What is known is that, when an alternating voltage is imposed at a certain frequency onto an antenna, radio waves are produced at that frequency, and travel outward at the speed of light. (In fact, light waves and radio waves are of the same nature, but of different frequency.)

Radio waves are produced by alternating currents above about ten to fifteen thousand cycles per second. When a radio wave strikes any electric conductor, it induces an alternating current in that conductor at the same frequency as the wave.

Long before any kind of radio waves were produced, they were known to exist. In fact, as early as 1865, James Clark Maxwell mathematically proved their existence. After Maxwell's startling

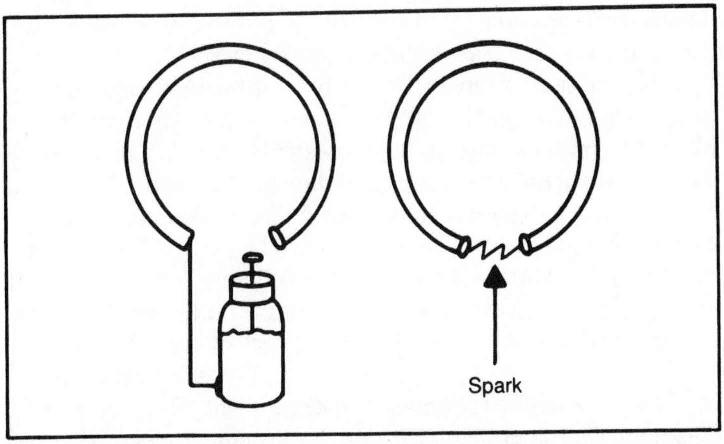


Fig. 8-1. This is the general idea of the apparatus used by Hertz to demonstrate the existence of radio waves. The Leyden jar was first charged, then discharged through the loop. On the other side of the room, another loop of similar size produced a spark at the instant the Leyden jar was discharged.

publication, however, over 22 years were to pass before physical proof was accomplished.

In 1887, Heinrich Hertz performed a carefully calculated experiment to prove Maxwell's discoveries. He discharged a Leyden jar through a loop of wire. Nearby, a similarly-sized loop with a gap in it produced a spark. See Fig. 8-1. In later experiments, Hertz was to prove that these mysterious waves were similar in nature to light waves—they could be focused, reflected, and refracted. He was, however, far ahead of his time. Some of his experiments used frequencies in the microwave region, and it was to be many years before those frequencies could be put to practical use.

## EARLY SYSTEMS

Within a decade, sensitive detectors were being produced, and Sir Oliver Lodge outlined a complete wireless system. In Russia, Popoff put a telegraph system into actual operation. The authorities felt that such communication was "subversive" and banned it. It wasn't until Marconi proved the commercial feasibility of radio telegraph that the system became popular.

One of the more important of Hertz's discoveries was that these mysterious waves could be tuned. At first, he didn't quite understand what was happening. What he did know was that the loops on the transmitter and on the receiver had to be about the same size, a condition that he called "syntony." In 1898, Lodge perfected the coil-and-capacitor tuning system, and three years later Marconi made his first transatlantic radio contact.

Marconi used a device similar to an automobile ignition coil to produce radio signals, and for a detector, he used a small tube filled with metal filings, called a *coherer*, Fig. 8-2. The principle was that, when radio signals passed through the filings, they would cohere, or stick together, making the internal resistance of the device much less. This would allow a local battery to close a relay, ringing a bell. At the same time as the bell rang, a vibrator would loosen the filings in the coherer, preparing it for the next radio impulse. It was, at best, a very inefficient device by today's standards.

Then came the crystal diode and DeForest's vacuum tube. Much of the early development and experimentation was done by amateurs. On the professional side, Fessenden produced the first voice-modulated broadcasts in 1900 and 1906. Radio operators all over the Eastern seaboard of the United States were astounded. One ship's radio operator entered into the log that he heard "angels singing."

Radio really came into its own, however, after World War I. In 1920, five American stations began commercial broadcasting, and within two years, there were over 500 on the air.

Since the invention of the vacuum tube and its related circuits, the means of producing radio waves has changed only slightly, except for the transition from vacuum tubes to transistors. Methods of modulation and very-high-frequency techniques have changed radically, while still using many of the earlier ideas.

The device that produces the high-frequency alternating cur-

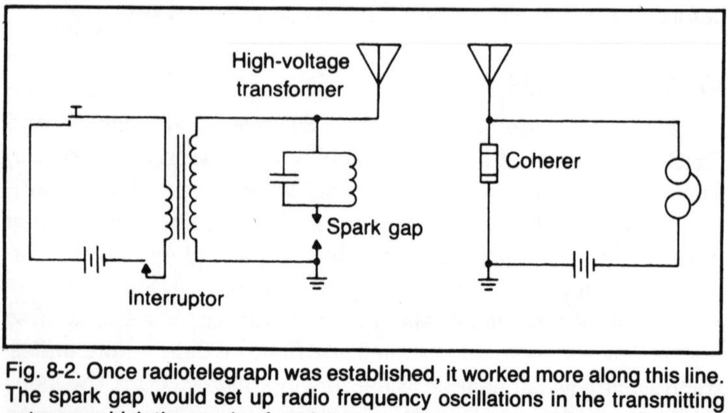


Fig. 8-2. Once radiotelegraph was established, it worked more along this line. The spark gap would set up radio frequency oscillations in the transmitting antenna, which the receiver's coherer would pick up. Not shown is a vibrator which would decohere the detector and get it ready for the next signal.

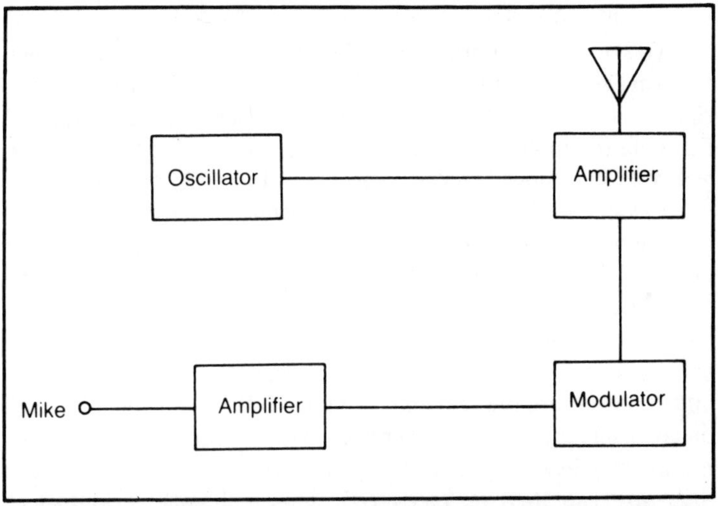


Fig. 8-3. Today's radio transmitter (at least, the AM transmitter of the last decade) consists of an oscillator which generates the radio signal, and an amplifier to give it power. The voice is first amplified, then fed to a modulator which puts it onto the radio signal.

rent which the antenna converts into radio waves is called an *oscillator*. In past years, an oscillator made from a high-power tube made a simple, one-tube transmitter. These were very popular among beginners on the amateur bands for code transmission. There were even some simple, one-tube voice transmitters, Fig. 8-3, but they simply don't make the grade for today's operating techniques.

## VOICES

The earliest, and for many years the main method of transmitting human voice involved the use of an audio amplifier equal in power to the output amplifier of the transmitter, which imposed the audio signal directly onto the radio-frequency output of the transmitter. This is still used in some CB transceivers. It results in what is called a *double-sideband AM signal.*

Sidebands result from the natural mixing of the audio and radio frequencies. They are caused by a law of nature that says that, when two signals of different frequencies are mixed together, they produce signals at the sums and the difference of the two. Applying that rule, if an audio frequency of 1000 Hz was mixed with a million hertz (1 MHz) of radio frequency (a hertz is one cycle per second), *sidebands* would be produced at 1,000 hertz above and below the one-million-hertz signal. That is, the sidebands would occur at 999,000 and 1,001,000 hertz. See Fig. 8-4.

You will notice from the illustration that an AM double-sideband signal consists of a carrier wave and the sidebands. If the signal is being modulated with a human voice, the sidebands will fluctuate in amplitude and frequency within the limits of the carrier plus and minus the voice spectrum. In a fully-modulated signal, the sidebands contain half the transmitter power and the carrier the other half. This is a tremendous waste. The only function of the carrier is to mix with the sidebands in the receiver to recover the voice.

Shortly before World War II, a means was developed to eliminate the carrier and one of the sidebands, putting the entire transmitter power into the remaining sideband. With this technique, all the available transmitter power is used to transmit the information. A carrier is inserted in the receiver at a much lower power, and the voice can then be recovered. This technique is known as suppressed-carrier, single-sideband transmission, Fig. 8-5. Besides using only half as much of the radio spectrum, it has the effect of a two- or three-to-one increase in power over an AM signal of the same power.

We have so far confined our discussion to transmitters, as they are generally the simpler devices in radio communication. At least, they were until single sideband communication came into its own. It is relatively easy to put a signal onto the air. Recovering that signal and sorting it out from among all the others is another matter entirely.

When a radio wave strikes a conductor, it induces minute alternating currents of the same frequency as the wave in the con-

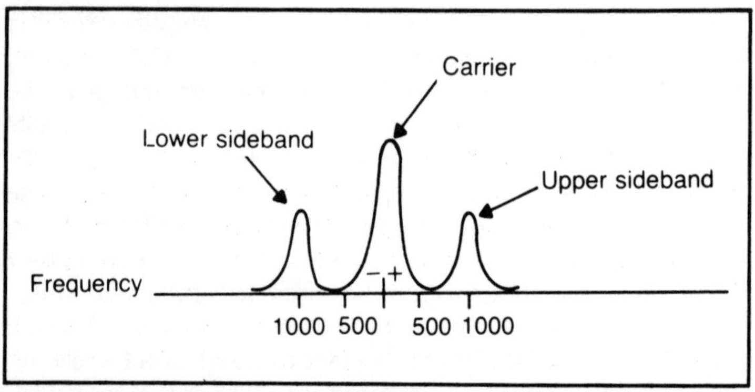


Fig. 8-4. If you made a graph of the frequencies being transmitted, you would find that a carrier wave and two sidebands are present.

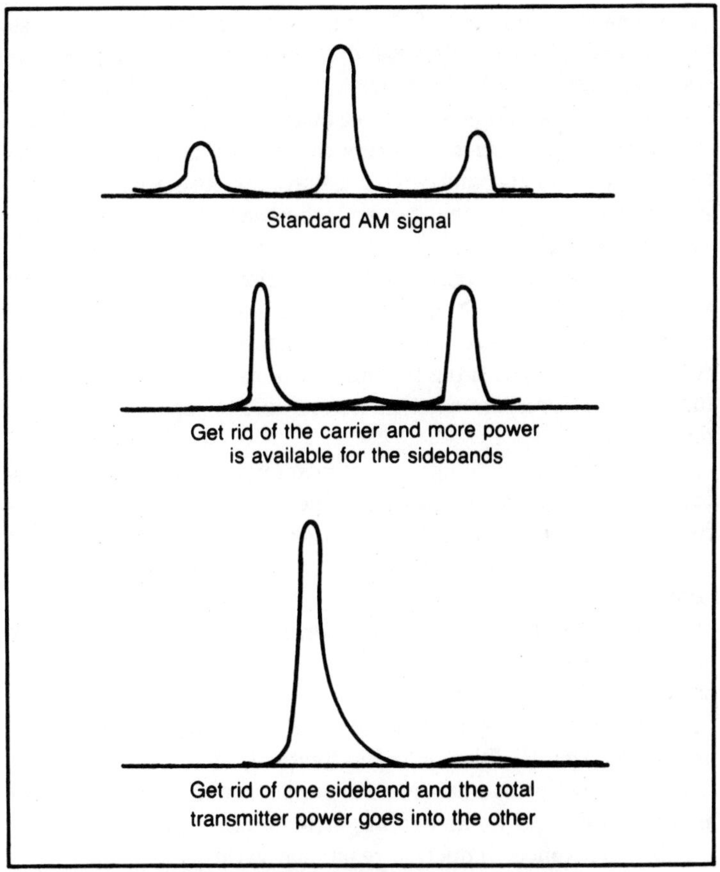
Standard AM signal
Get rid of the carrier and more power is available for the sidebands
Get rid of one sideband and the total transmitter power goes into the other
Fig. 8-5. The evolution of single sideband signals.

ductor. Now, there are thousands of different waves striking every antenna constantly. Before detecting them, we must first sort out the wanted one.

## RESONANCE

In order to better understand the process of electrically sorting signals of different frequencies, we need first understand the mechanical parallel to the phenomenon. Every object that is made to vibrate has what is called a natural period; that is, a frequency at which it vibrates much more easily than any other. That frequency is referred to as the *resonant period* of the object, and is affected by a number of factors.

A good example is the string in a musical instrument. Its natural resonant period depends on such factors as length, thickness, elasticity, and tension. Vary any of these, and the frequency at which the string will vibrate changes.

Resonances were long a major bugaboo in the development of microphones, earphones, speakers, and all devices associated with the reproduction of sound. As the art progressed, scientists learned how to avoid those resonances, and when unavoidable, to use them to an advantage.

All this has been paralleled in electrical considerations. Just as mechanical devices have mechanical resonances, electrical devices have natural resonances at which frequencies alternating current functions with the maximum efficiency. It is by the careful use of such resonances that we are able to separate signals in receivers.

The two factors having the greatest influence on electrical resonance are inductance and capacitance. Except at the very high frequencies, these two factors are relatively easy to control, not only through good layout but through use of the correct components as well.

The handiest thing about inductors and capacitors is that they are electrically opposite to one another. Certain factors in their opposition to the flow of alternating current (a factor dependent on frequency) are positive in an inductor and negative (relatively speaking) in a capacitor. Thus one can effectively cancel out the other.

The inductive and capacitance factor that opposes the flow of alternating current (to an extent partly determined by frequency) is called reactance. Inductive reactance varies directly with frequency; capacitive reactance varies inversely with frequency. That is, as frequency increases, inductive reactance becomes greater and capacitive reactance becomes less. See Fig. 8-6.

It stands to reason, then, that in a circuit having both inductive and capacitive components, there must be a frequency at which inductive and capacitive reactances are equal. At that frequency, and that frequency alone, the two cancel one another out entirely and the only opposition to the signal is the pure resistance. Therefore, the circuit will show a far greater efficiency at that frequency than at any other. It is said to be *resonant* to that frequency.

Let's look at that a little more closely. If a coil and a capacitor were connected in series, Fig. 8-7, the coil would offer greater opposition to high frequencies than at low frequencies. The exact amount of opposition would depend on its inductance as well as on the frequency. The capacitor would offer greater opposition to

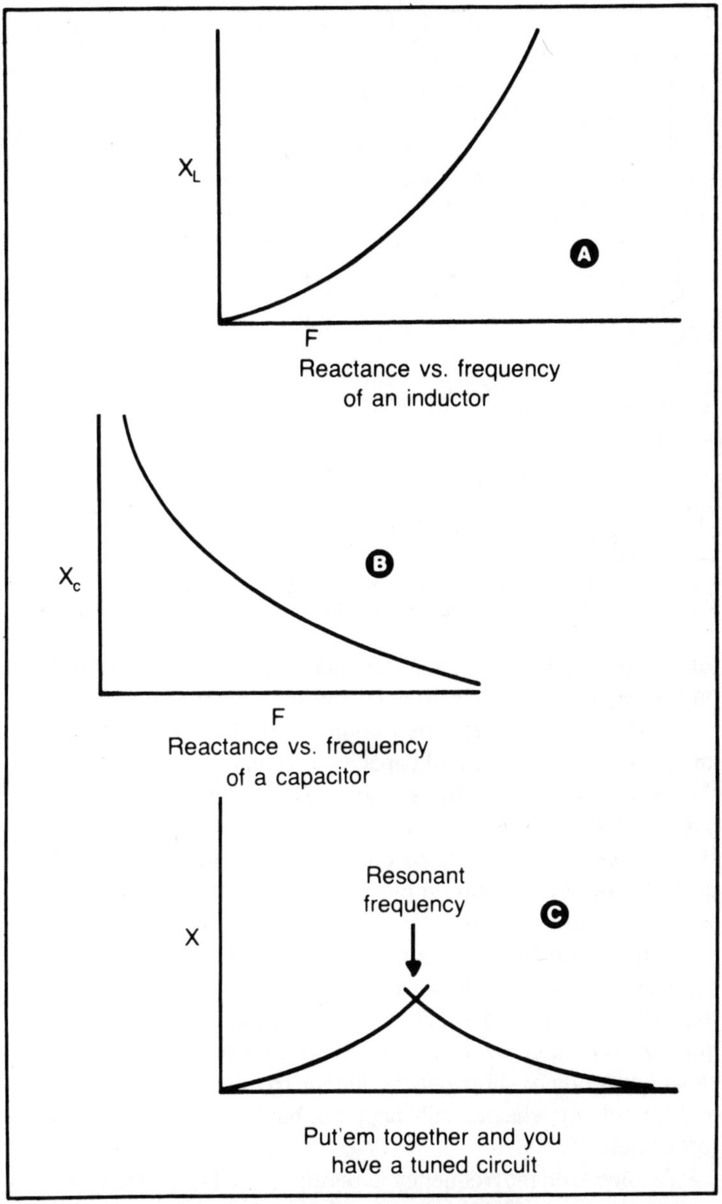


Fig. 8-6. The reactance of an inductor varies with frequency (A). The higher the frequency, the higher the reactance. With a capacitor, it's the other way around. This means that there is one particular frequency that can flow best through an inductance/capacitance combination. That frequency is called the *resonant frequency*. At resonance, the two reactances cancel one another (C), and only resistance is left to be dealt with.

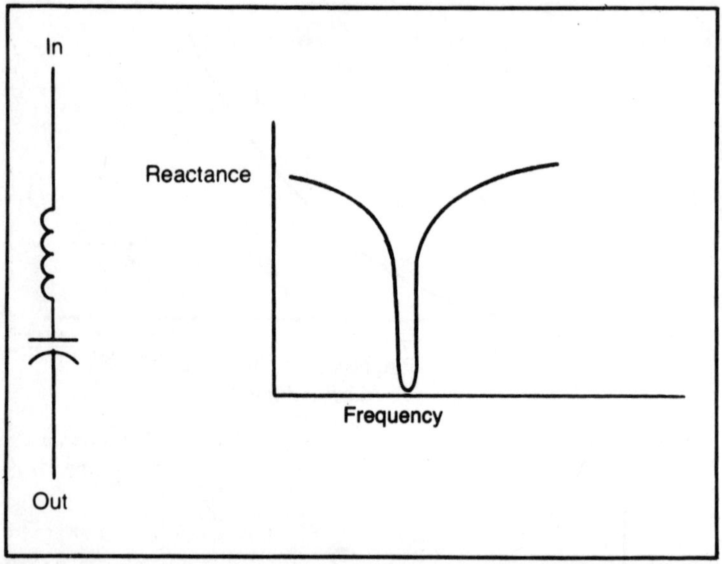


Fig. 8-7. The reactances of a series tuned circuit roughly follows this curve. A series tuned circuit lets the resonant frequency pass through easily.

lower frequencies. The exact amount of opposition would depend on the capacitance as well as on the frequency.

If the values of the two components are fixed at specific amounts of inductance and capacitance and remain unchanged as the frequency is varied, there would be a point where one cancelled out the other and only the pure resistance of the wire would be left. Higher frequencies would be opposed by the inductor; lower frequencies would be opposed by the capacitor. This point is known as the natural resonant frequency of the circuit.

Now picture the two components in parallel with one another, connected across the signal source, Fig. 8-8. Again let us assume that the values are so selected as to be resonant at the signal frequency. The inductor tends to short out the lower frequencies; the capacitor tends to short out the higher frequencies. This is really oversimplifying what actually happens, but will do for what we have to say here.

At the resonant frequency, a parallel tuned circuit offers maximum opposition. There are other factors, much too advanced to go into here, but let us simply say that a parallel tuned circuit also emphasizes its resonant frequency, much like an organ pipe does with sound. A parallel tuned circuit, connected across a signal source, will actually show an apparent increase in the signal volt-

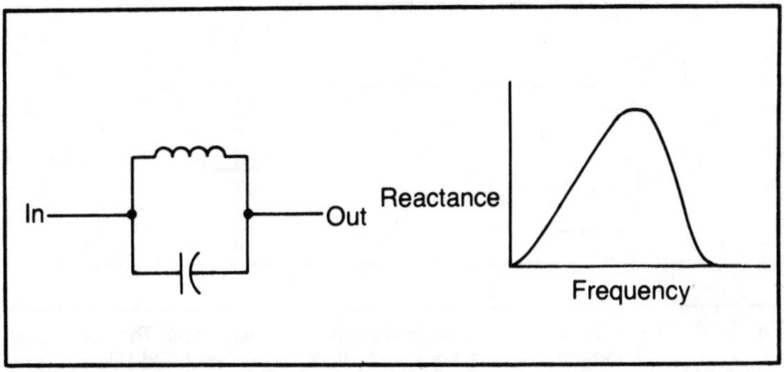


Fig. 8-8. A parallel tuned circuit behaves in an opposite manner than a series circuit; the resonant frequency is blocked.

age. It is, therefore, the type most often used in radio detection. In most AM radios, the capacitor is varied to select the resonant frequency.

This lengthy discussion of resonance has shown the way in which incoming signals are separated from one another. Once this is done, the next thing to do is to recover the information from the selected signal. A simple tuned system is connected as shown in Fig. 8-9.

Once the desired signal is isolated by the tuned circuit, information can be recovered by rectifying it in the case of an AM signal, or by mixing it with a carrier in the case of a single-sideband or Morse-code signal. (FM signals are in an entirely different ball game and the means of detection are a bit too complex for this book.)

Fig. 8-9. The basic antenna/ground circuit of a radio system uses a parallel tuned circuit. All unwanted frequencies are shorted to ground, and the desired frequency is left to be processed.

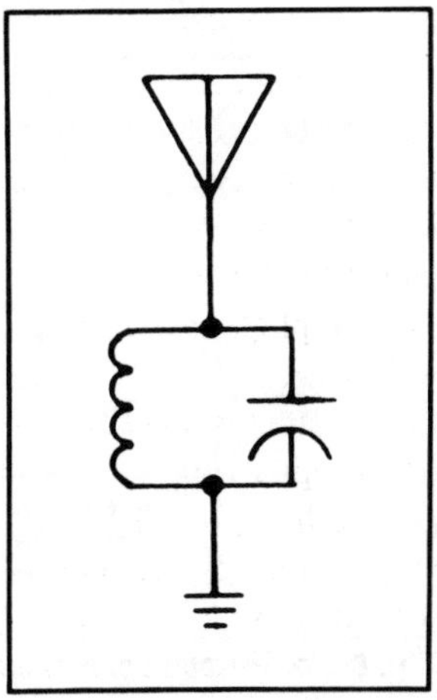

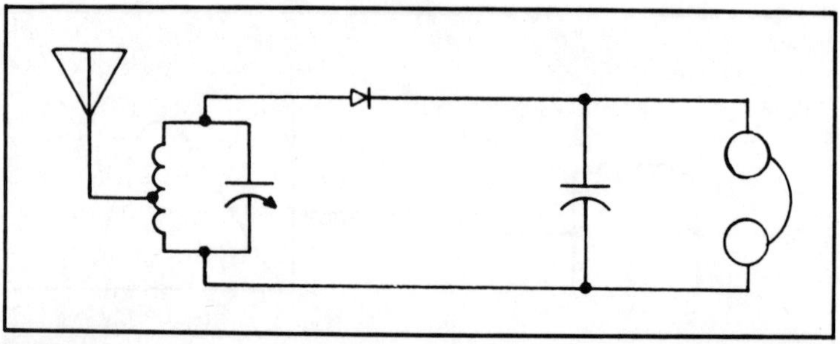

Fig. 8-10. The old crystal set of Grandpa's age used this circuit. The catwhisker detector was connected where we use a diode. You can build this set, and it will work. The variable capacitor has a value of 365 pF. The coil is No. 28 wire wound on a paper roller about 1 1/4 inch wide for a length of five inches. The fixed capacitor is 0.001 $\mu$F.

## RECEIVERS

The earliest receivers (once the coherer went out of style) consisted of a galena crystal with a thin wire called a *catwhisker* touching it. This arrangement acted like a diode. Substituting a diode for the crystal detector, we end up with the circuit shown in Fig. 8-10.

When the carrier and sidebands are selected by the tuned circuit, they then mix in the diode to form a radio-frequency signal with an amplitude that varies at an audio rate. This is then rectified by the diode to produce a dc voltage that fluctuates at an audio rate. This voltage is then fed into the earphone where it produces sound. The circuit is shown in Fig. 8-11.

The earliest transmissions were, of course, in code. However, they were still detectable with a diode detector because the trans-

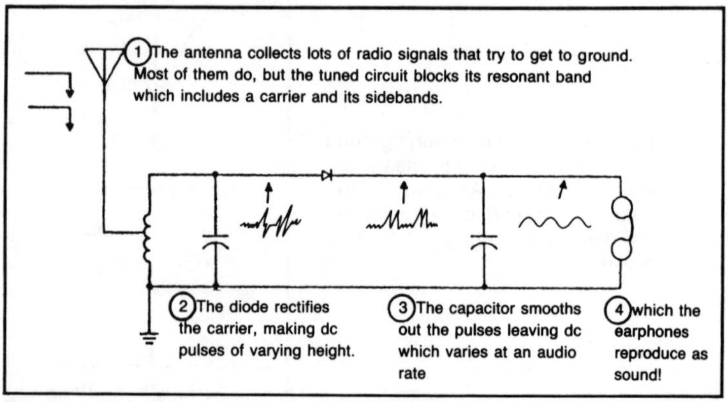

1 The antenna collects lots of radio signals that try to get to ground. Most of them do, but the tuned circuit blocks its resonant band which includes a carrier and its sidebands.
2 The diode rectifies the carrier, making dc pulses of varying height.
3 The capacitor smooths out the pulses leaving dc which varies at an audio rate
4 which the earphones reproduce as sound!

Fig. 8-11. Operation of a simple radio receiver.

mitter was a spark gap. The spark produced an interrupted signal which, once rectified, made a buzz in the earphone. A pure, steady-state signal would only have produced a tiny dc voltage, and nothing more than a click would have been heard in the earphone.

The earliest receivers were crystal detectors. Even the best of them had a tuned circuit that was primitive by today's standards. Even though there were relatively few stations on the air, interference was very bad. Perhaps that is why many of the early developments that improved the art of radio communication and helped prove the commercially developed circuits were accomplished by amateur radio experimenters. They had little choice but to get along with what they had, and to get the most out of it.

During the babyhood of radio, the typical amateur transmitter consisted of a tuned spark gap. Simple as that may sound, it wasn't too far removed from the commercial gear. The receiver was simply a crystal detector. Figure 8-12 shows a typical one that was in the 1916 Boy Scout Handbook.

Please take my word for it that this worked, at least over relatively short distances. I do *not* advise any experimenter to build

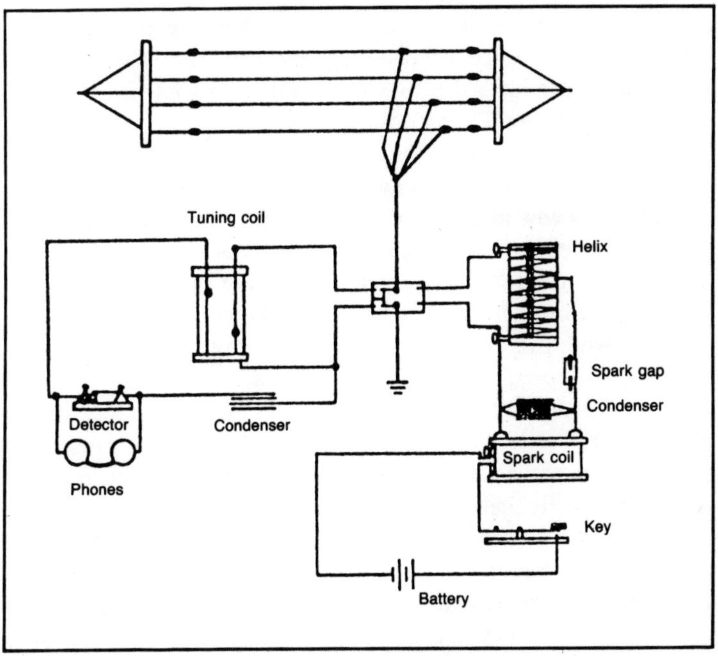


Fig. 8-12. This is a copy of the circuit of a two-way radio telegraph project in the 1916 Boy Scout Handbook. Do not try to put this on the air!

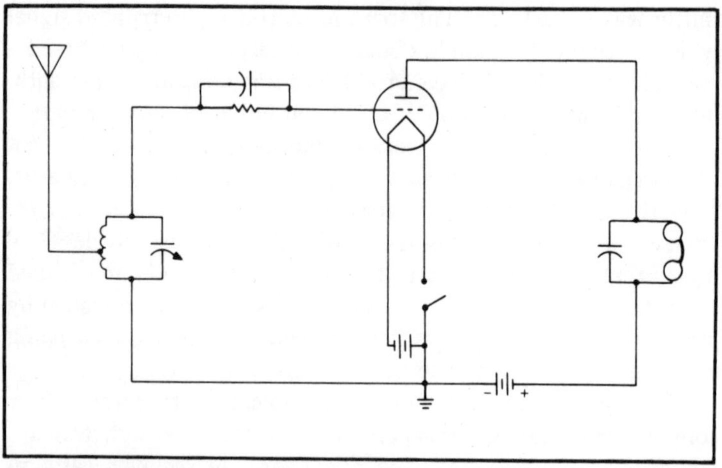

Fig. 8-13. The basic grid-leak detector. Tubes in those days had no separate cathode, and the circuit was wired as shown.

and operate it. While it did create a radio signal, that signal was filled with noise and harmonics. If you put a spark transmitter on the air today, you would have every amateur, commercial, TV, and government radio station for miles around howling at your door—not to mention a neighborhood lynch mob and the FCC. You see, the signals it produces are not clean, so there is some interference to everything from TV on down.

With the invention of the vacuum tube, the improvements in electronics grew tremendously. Hardly a year went by without a revolutionary new innovation. Vacuum-tube detectors applied the signal to the grid of the tube, and the grid was biased so as to rectify. That is, the tube only conducted during the positive half of the cycle. Consequently, the signal not only was detected, but amplified as well. Figure 8-13 shows the circuit of an early triode detector.

This detector was known as a *grid-leak detector* because of the resistance in series with the grid. The signal passed through the capacitor, but the dc charge that built up naturally on the grid was allowed to slowly leak off through the resistor. This kept the grid biased slightly negative so that signals could be rectified. If you want to build this on your experimenter breadboard, the values given in the diagram will work. In the interest of space, we will give you the circuits without going into elaborate instructions, except as noted in the illustrations.

The setup with which DeForest first demonstrated the poten-

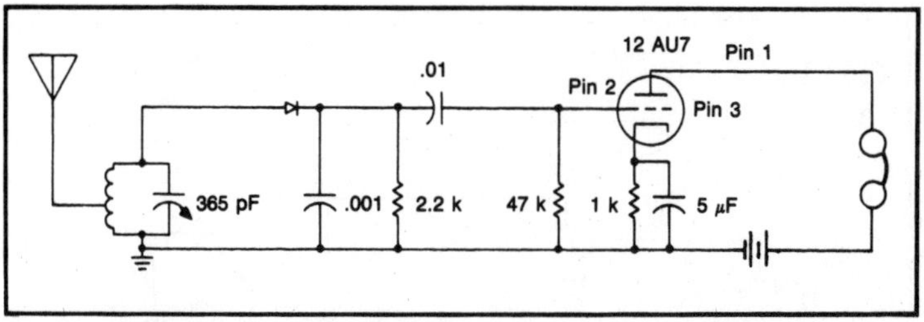

**Fig. 8-14. Use the diode detector and an amplifier circuit to make this version of the first amplified radio receiver.**

tial of the vacuum tube was probably a crystal detector with an audio amplifier coupled to it. A modern-day version of that setup is shown in Fig. 8-14.

The difference between this circuit and DeForest's is that the early versions used a transformer to couple the amplifier stage to the detector. If you add a couple of switch sections as shown in Fig. 8-15, you can then switch the amplifier in and out, demonstrating the amplifying ability of the tube.

Then somebody had the idea of adding an amplifier stage before the detector (Fig. 8-16). Since there were now two tuned circuits instead of one, the system had much sharper tuning, but there were some disadvantages.

A big breakthrough came when Major Edwin Armstrong discovered a method of feeding some of the amplified signal back to the input circuit. This brought about tremendous increases in gain, and made very sensitive, simple sets a reality. It was called the *regenerative detector*. See Fig. 8-17.

Up until World War II, the average amateur receiver consisted of a regenerative detector with one or two stages of audio amplifi-

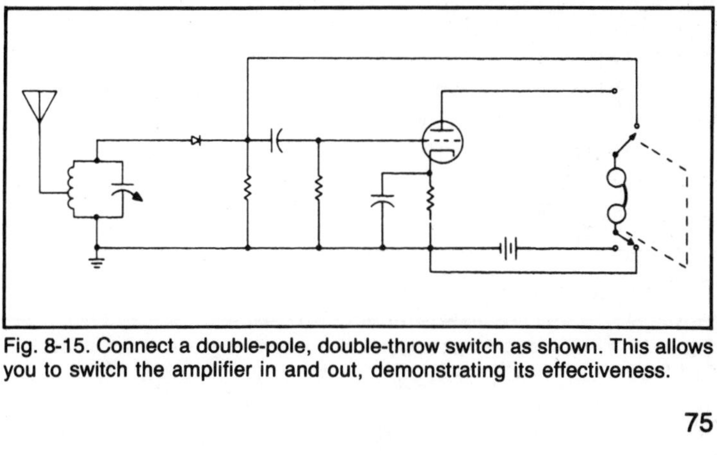
Fig. 8-15. Connect a double-pole, double-throw switch as shown. This allows you to switch the amplifier in and out, demonstrating its effectiveness.

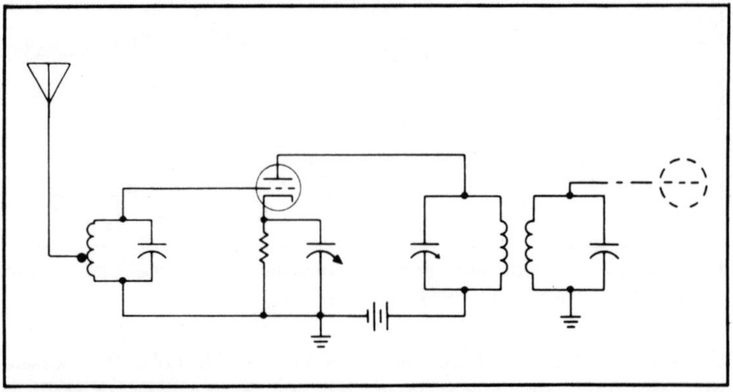

Fig. 8-16. A radio-frequency amplifier. This circuit is incomplete, and if you build it, it may oscillate.

cation. Then, shortly before the war, Major Armstrong devised another circuit called the superheterodyne. It operates on the principle that, when two signals of differing frequencies are mixed together, a new signal is produced. Actually, there are two new signals. Their frequencies are the sum and the difference of the original two.

By using the new frequency that is lower than the original frequency, it is possible to use circuits that tune much sharper. Therefore, the whole radio had much sharper tuning, as well as being more sensitive than even the regenerative detector.

Let's look at the way a superheterodyne works. Instead of showing all the components, we'll represent each stage of the system as a box. Such a diagram is known as a *block diagram*. See Fig. 8-18.

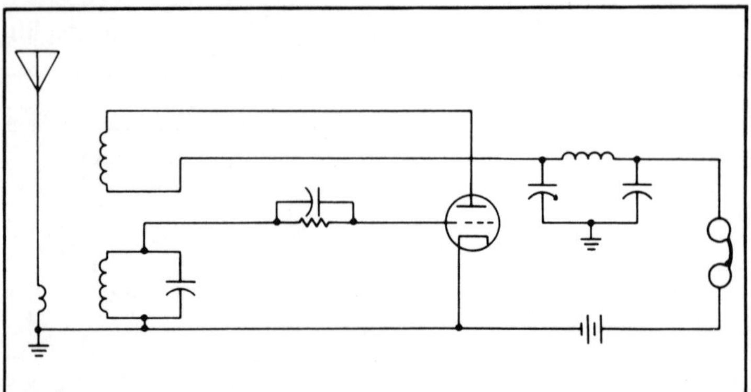

Fig. 8-17. Edwin Armstrong's regenerative receiver was a major innovation that became the favorite for amateur receivers prior to World War II.

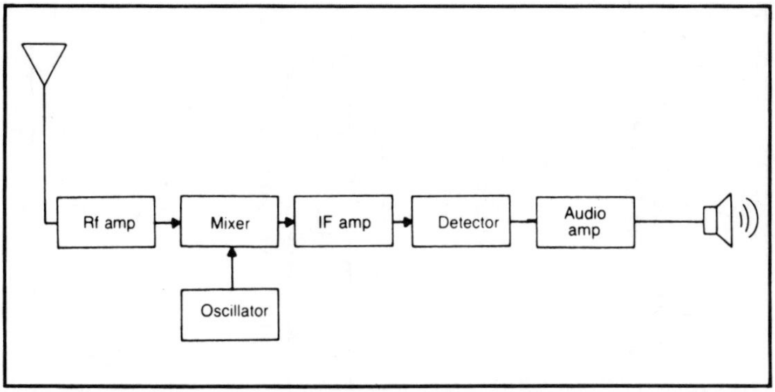

**Fig. 8-18. Armstrong later came up with the superheterodyne circuit that is still the basis for radio receivers today.**

The signal may, depending on how expensive a receiver it is, pass through an amplifier (called an *rf amplifier* because it amplifies radio frequencies) into the signal mixer. There it combines with a signal from an oscillator to produce the intermediate frequency (I-F). This frequency (very often 455 kHz) is the amplified. From there, it passes to a diode detector, then to the audio amplifier. See Fig. 8-19.

## TRANSMITTERS

Now we turn our attention to transmitters, and go back to the old Armstrong regenerative receiver (Fig. 8-16). If too much energy were fed back to the input, the circuit would begin to oscillate on its own. When this happened, the signal produced was much steadier than anything produced by even the finest spark transmitter. However, in order to be received, the detector receiving it had to be oscillating also. It was tuned slightly off the transmitted frequency so that the difference between the received signal and the oscillation of the detector was within the audible range. This produced the pure, musical note that we now identify with radio telegraph dots and dashes.

The modern superheterodyne receiver has a built-in oscillator that produces this note when it receives code signals, and provides a carrier for signal-sideband voice reception.

Sometime between the two world wars, quartz crystals came into use for frequency control. When an electric signal is imposed on a section of quartz crystal, it produces a mechanical strain. The instant that strain is released, the crystal vibrates for a few micro-

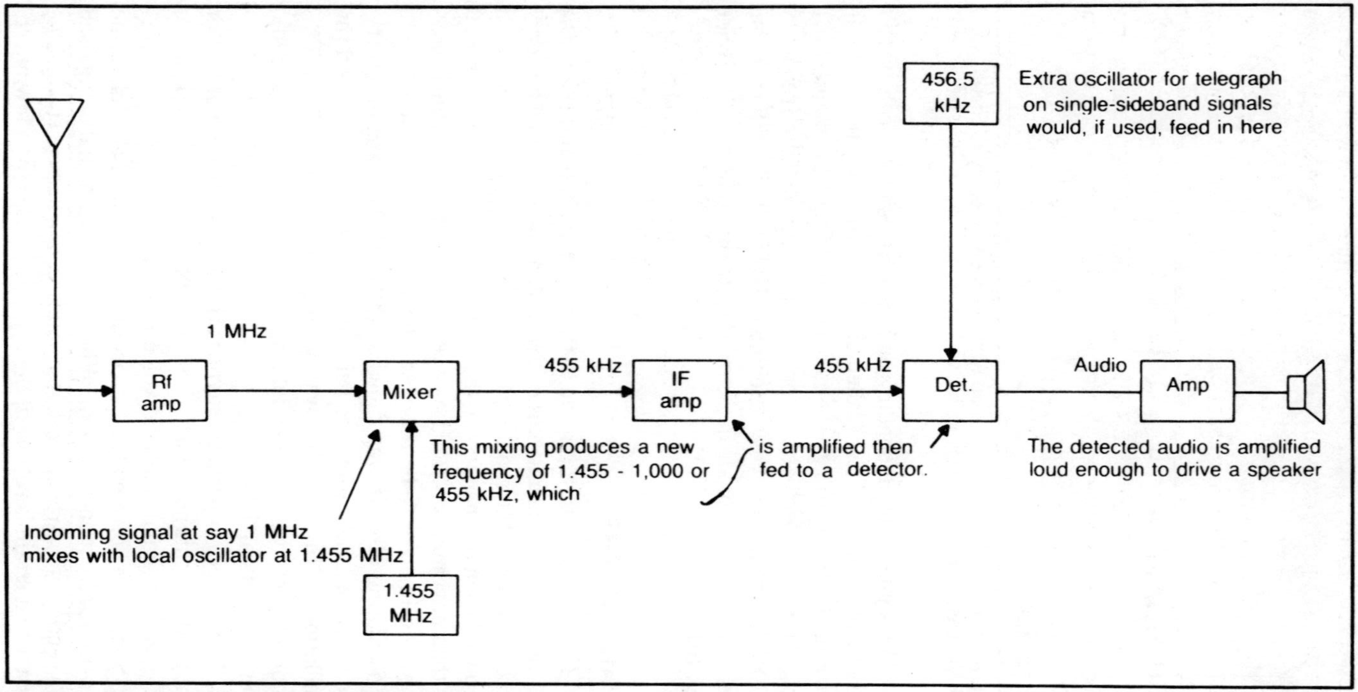
456.5
kHz
Extra oscillator for telegraph
on single-sideband signals
would, if used, feed in here
1 MHz
Rf
amp
Mixer
455 kHz
IF
amp
455 kHz
Det.
Audio
Amp
This mixing produces a new
frequency of 1.455 - 1,000 or
455 kHz, which
is amplified then
fed to a detector.
The detected audio is amplified
loud enough to drive a speaker
Incoming signal at say 1 MHz
mixes with local oscillator at 1.455 MHz
1.455
MHz

Fig. 8-19. Operation of a superheterodyne circuit.

seconds and that vibration causes it to deliver a minute alternating current at the frequency of the vibration. This frequency is dependent upon the mechanical dimensions of the crystal fragment. Therefore, if the crystal is placed anywhere in the feedback circuit of an oscillator, it will control the oscillator frequency. Figure 8-20A shows an oscillator circuit and its crystal-controlled equivalents, Fig. 8-20B.

Until the introduction of transistors, the standard transmitter arrangement consisted of an oscillator followed by one or two amplifier stages. Some were crystal controlled; others used a variable-frequency oscillator (VFO).

Voice transmission was achieved by simply amplifying the voice to a power level equal to the transmitter output power, and connecting it through a transformer in series with the output amplifier, Fig. 8-21. It was an expensive way to do it but, for a long time, the only practical way.

Transmitters now are quite different. For one thing, they use transistors instead of vacuum tubes (with the possible exception of the output stage). Also, instead of amplitude modulation (except in broadcasting and CB), they produce a single-sideband signal.

The signal is generated at a relatively low frequency and mixed with the audio signal in what is called a balanced modulator. This produces two sidebands with no carrier. One of the sidebands is filtered out, leaving a low-frequency, single-sideband signal. This signal is then mixed with the signal from another oscillator to produce a higher-frequency signal in the same manner as two signals

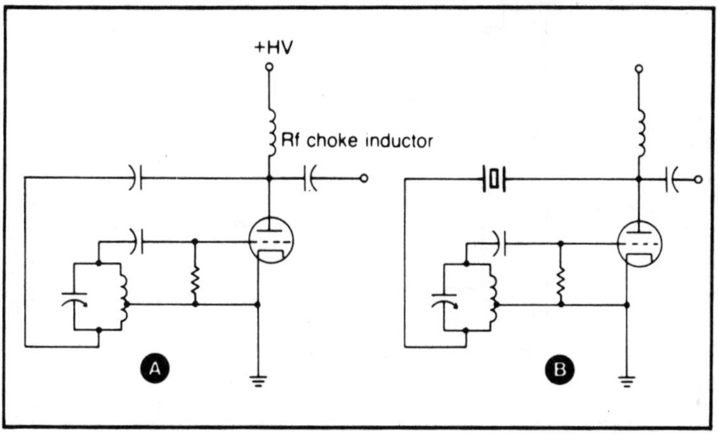


Fig. 8-20. How a crystal can be added to an oscillator circuit to control its frequency.

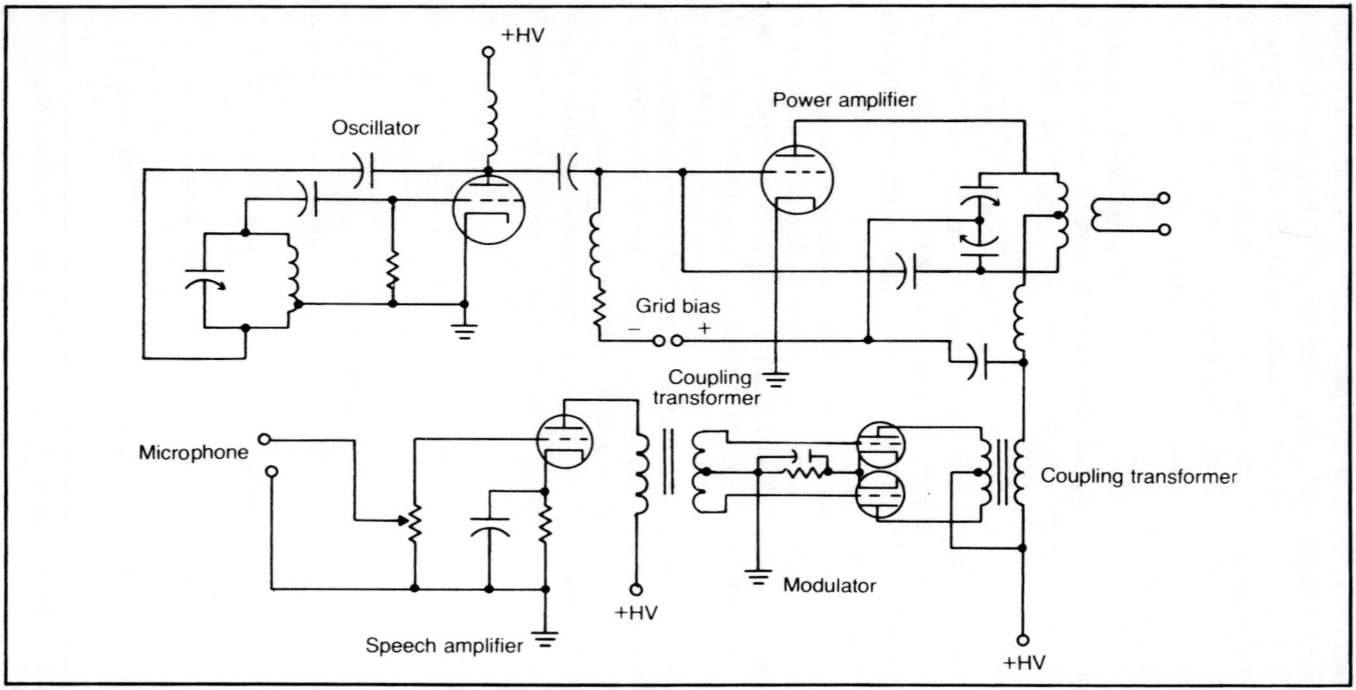


Fig. 8-21. The combination of circuits that make an AM radio transmitter.

are mixed in a superheterodyne receiver. When the output signal is at the proper frequency, it is amplified and applied to an antenna. Figure 8-22 is a block diagram of a single-sideband transmitter.

It would be possible to write an entire book on radio transmitters and receivers, but what we have covered is more than enough for one chapter. For further study, there are a number of excellent TAB books on the subject.

## TELEVISION

Television uses all the techniques of radio communication, along with photographic techniques and a couple of other instruments. There are two transmitting systems used. One is a frequency-modulation (FM) transmitter no different from those used in FM broadcasting. This carries only the sound channel. The other produces a single-sideband AM signal with the carrier present. This carries all the picture information. See Fig. 8-23. The picture signal contains so much information to be transmitted so quickly (30 complete pictures per second) that it occupies six times as much of the radio spectrum as the entire AM broadcast band.

The picture signal is very complex. It contains the actual information of the picture elements, vertical and horizontal synchronizing information, and color information. To better understand all this, let's look at it a bit at a time.

First the actual picture information. The TV camera focuses the picture on a photo-sensitive screen which develops an electric charge that is proportional to the amount of light, and precisely distributed as an electrostatic image. This screen is inside a special kind of vacuum tube. The electron beam from the cathode of the camera tube is caused to sweep back and forth across the picture. This produces a varying dc output that is a precise electrical record of the light and dark parts of the picture as they were scanned by the beam in the camera tube.

The video signal from the camera modulates an AM transmitter. At the receiver, the signal is detected, amplified, and fed to the grid of the picture tube. The picture tube produces a cathode ray similar to that in the camera tube, which sweeps back and forth across a fluorescent screen. This causes the screen to light up, and the brightness is proportional to the intensity of the beam. Since the beam is modulated by the video signal, the picture tube produces light and dark areas that are exact duplicates of the light and dark areas in the camera tube.

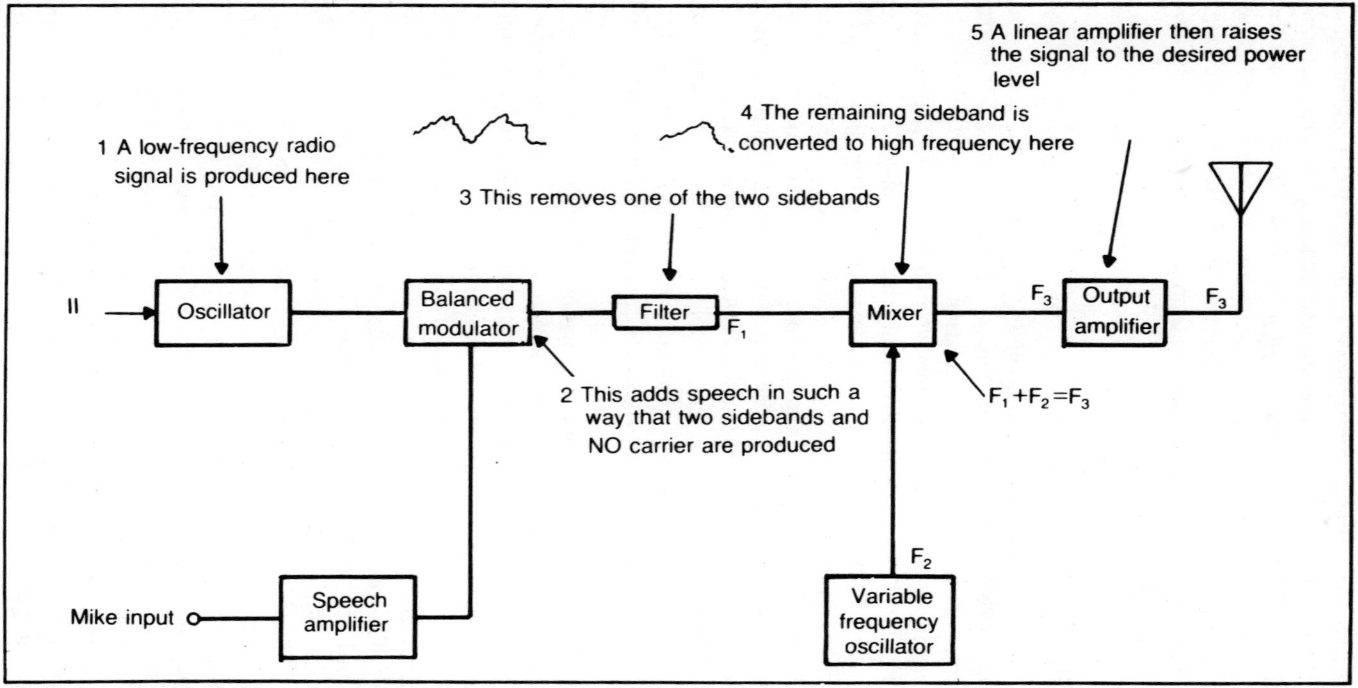


Fig. 8-22. Block diagram of a single sideband radio transmitter.

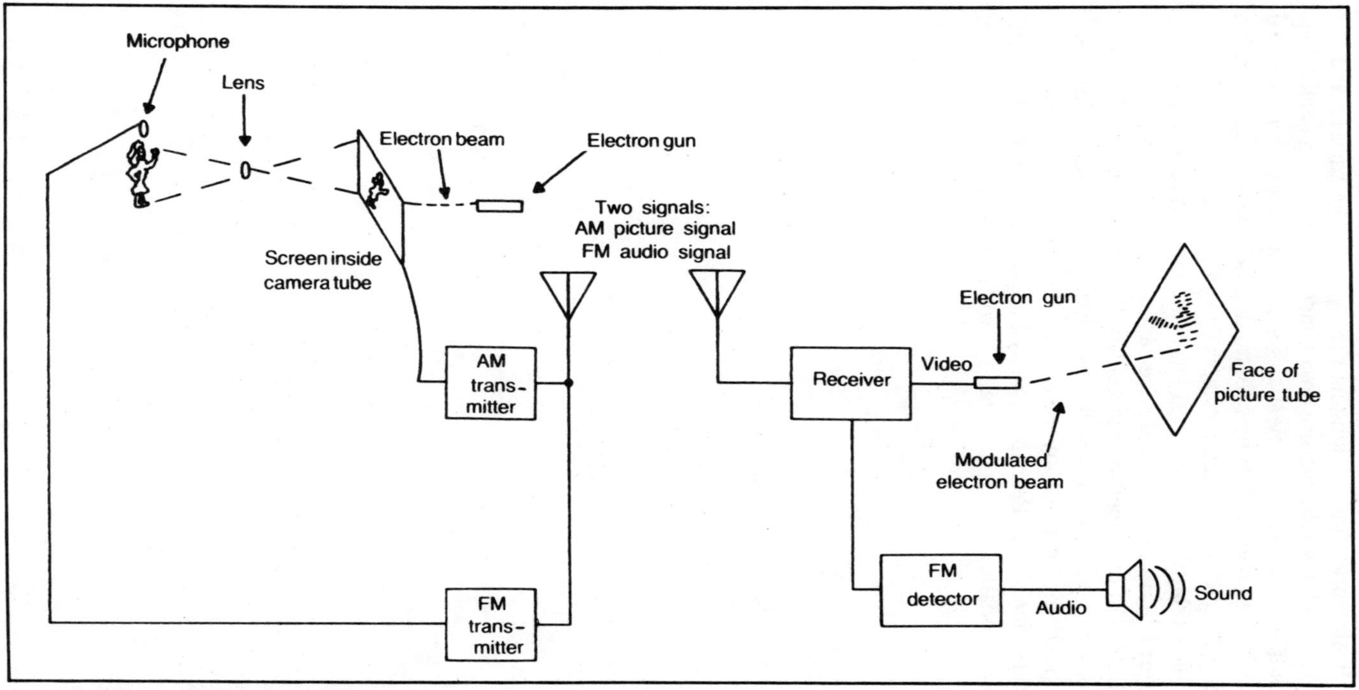


Fig. 8-23 The principles of a television system.

It all works fine, as long as the tubes in the camera and the TV receiver scan the picture area exactly in step with each other. This is accomplished by means of synchronizing pulses imposed on the video signal. A signal containing both picture information and sync pulses is called a *composite video signal.*

The television receiver, then, is a combination of an AM receiver, FM receiver, and a device called a cathode-ray oscillograph (the circuits that operate and control the picture tube). If color is involved, the complexity of the system triples. There are three sets of video information involved, together with a special color-information signal. Perhaps, when you realize all this, you may feel a bit sympathetic to your local TV repairman.

# Chapter 9
# Basic Rules of Wave Propagation and Electronic Tuning

EXCEPT FOR WAVELENGTH, RADIO WAVES ARE SIMILAR TO light waves. They can be focused, reflected, refracted, and polarized. Because of the difference in wavelength, radio waves pass easily through some substances that block light.

## WAVE PROPAGATION

A radio wave consists of an electric field and a magnetic field at right angles to one another as shown in Fig. 9-1. As the wave travels, the two fields change direction and intensity while maintaining their relationships to one another. Should these moving fields intercept an electric conductor, they would induce an alternating current in that conductor with a magnitude and polarity corresponding exactly to the magnitude and direction of the fields comprising the wave.

Radio waves travel at the speed of light: 299,793,077 meters per second, or 186,282.386 miles per second, in empty space. For convenience, we round off these figures to 300,000,000 meters and 186,000 miles per second. The distance a wave travels while making one complete cycle is called the wavelength. A complete cycle occurs when the two fields have undergone two complete reversals and returned to their starting point. Figure 9-2 shows a waveform with a cycle measured at a zero volt reference and at a positive voltage maximum. The number of waves passing in one second is

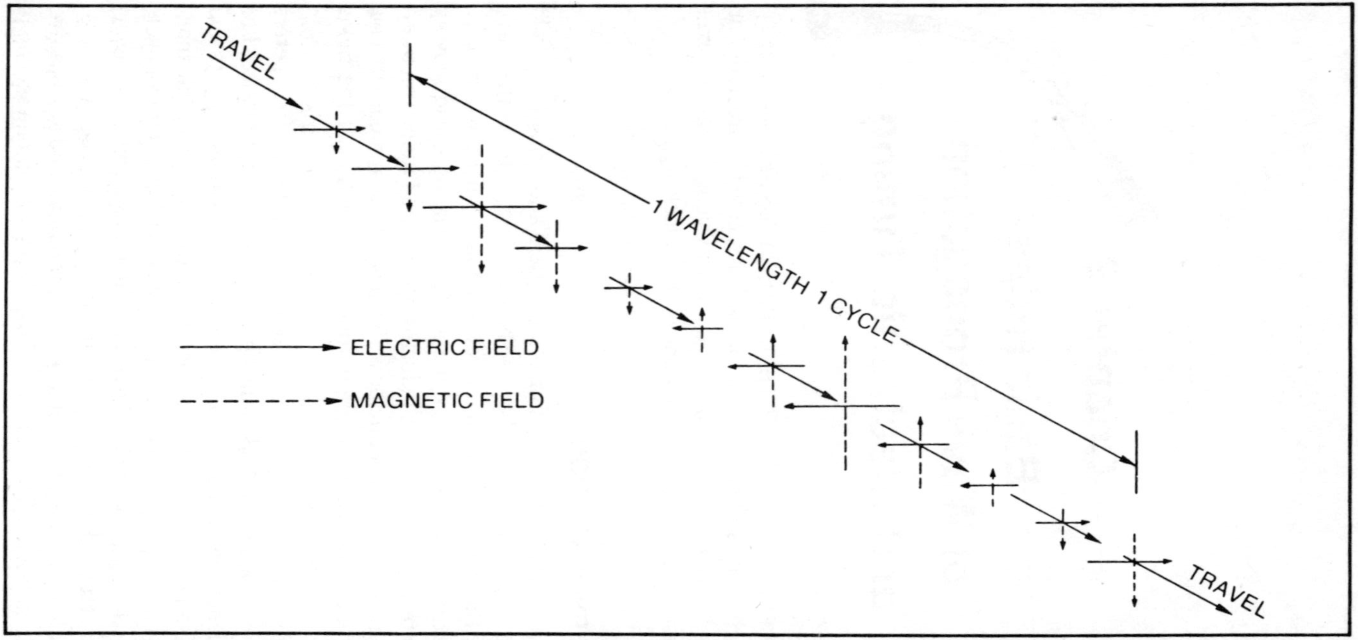


Fig. 9-1. Composition of a radio wave. The electric field is represented by the solid arrows, while the magnetic field is represented by the dotted arrows. The length of the arrows indicates the relative intensity of the two fields through various parts of a cycle. One complete cycle can be measured from any portion of the wave so long as measurement is made from the starting point to the next identical portion.

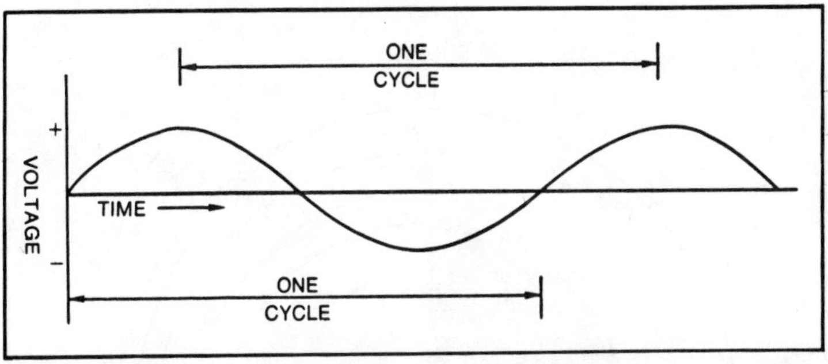


Fig. 9-2. Voltage wave set up in an antenna by a passing radio wave. Again note that one cycle is the time between any two points at which the magnitude and polarity of the voltage is identical.

called the *frequency*, expressed in hertz, kilohertz, or megahertz. A hertz is one cycle per second; a kilohertz one thousand; and a megahertz one million cycles per second.

Light travels in a straight line. If you light a candle, the base remains in shadow unless something reflects a light beam there. Similarly, a radio signal originating at any point on the earth's surface will not reach beyond the horizon unless something reflects it. Fortunately, nature has provided a reflector for high-frequency radio communication.

High above the surface of the earth are several layers of electrically charged particles collectively known as the *ionosphere*. Figure 9-3 shows the layers of interest to amateurs. These layers absorb waves of some frequencies and reflect waves of other frequencies. Variations in the particular frequencies affected and in the direction of reflection cause the various bands below 30 MHz to "open up" to different parts of the world from time to time.

Radio waves striking the ionosphere are reflected back to earth far from their point of origin (see Fig. 9-4). They can then be reflected back to the ionosphere and thence to earth again several times. The area between these skips, known as the *skip zone*, contains little or no signal.

One ordinarily would expect the signal to be radiated uniformly at all angles, thereby covering the entire surface with signal. Unfortunately this doesn't happen. The reflecting ability of the ionosphere is not distributed evenly thus it supplies concentrated signals to some places while leaving others void of signals. Also, because of the reflectivity of the ground, antennas tend to bounce their radiation at particular angles. High-angle radiation is best for communications from a few hundred miles to a thousand miles or so, while

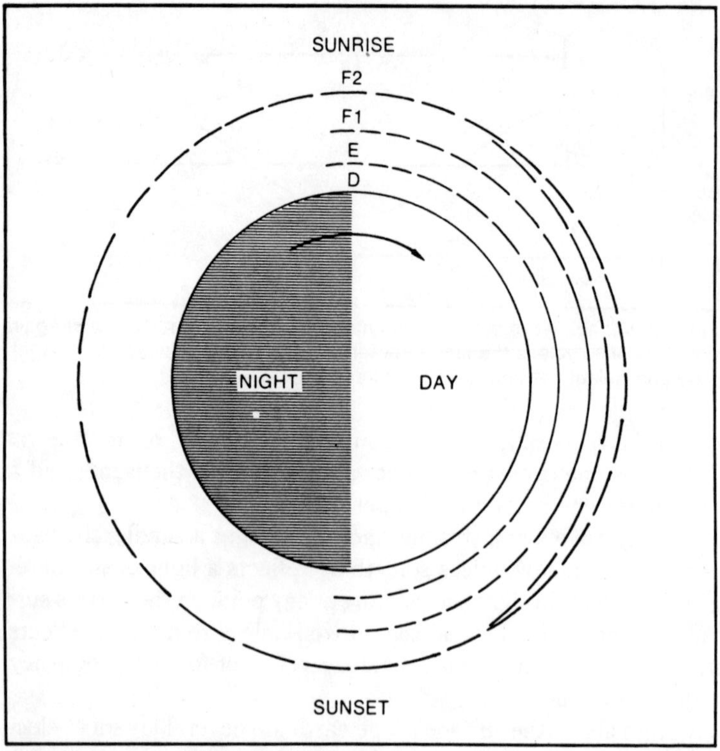


Fig. 9-3. The layers of the ionosphere. Continuity of the lines indicates the relative density of the layers. Note that only the F layer is present for a full 24 hours.

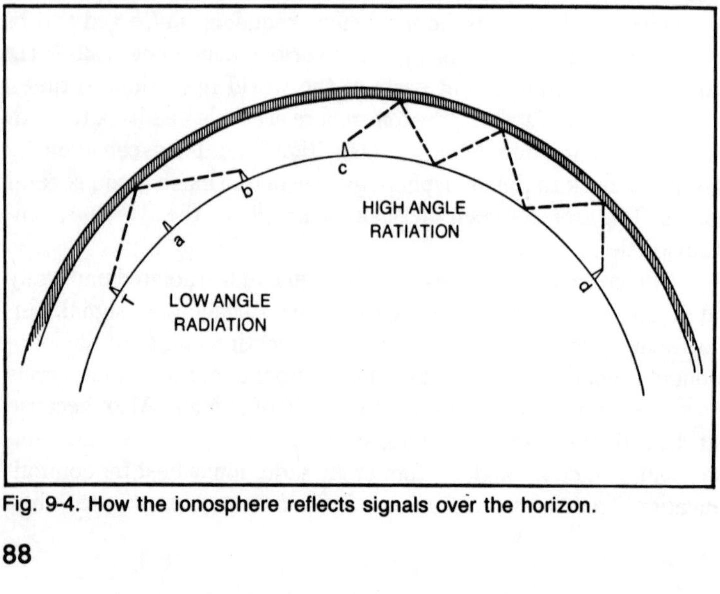


Fig. 9-4. How the ionosphere reflects signals over the horizon.

low-angle radiation brings in remote places.

A portion of the signal, known as the *ground wave*, follows the surface of the earth to just beyond the visible horizon. Beyond that range, if there are no signals being reflected back from the ionosphere, is a dead zone. Thus it is possible to hear stations hundreds of miles away while stations just a few miles away are received very poorly.

There are four layers of the ionosphere of particular concern to amateurs, labeled D, E, F1, and F2. The D layer, which only exists in the daytime, does little reflecting but absorbs signals, especially around midday when ionization is greatest. It is the nightly disappearance of this layer that makes the 80- and 40-meter bands open up at night. The E layer is responsible for some medium-distance propagation, but when ionization is high it can also absorb signals. It thins out just after sunset and makes a fast comeback the next morning. The F layer is the most stable of all, owing its stability to its height. It remains effective at night, although moving to a greater height, becomes weakest just before sunrise, and then stages a fast comeback. During periods of high ionization, it divides into the F1 and F2 layers. The lower of the two, the F1 layer, is similar to the E layer in its behavior, and does some absorbing. The F2 layer remains a useful reflector for daytime communications.

The particular frequencies absorbed or reflected, and the extent of absorption or reflection, is governed by the amount of ionization of the various layers. Ionization is caused by solar radiation, and consequently long-distance communications respond markedly to the 11-year sunspot cycle. The 20-, 15-, and 10-meter bands are comparatively dead during sunspot lull periods but distant stations come rolling in on these bands during peak activity periods.

From all this, we can see that, for long-distance communication, there exists upper and lower usable frequencies for communicating with various parts of the world. These frequencies vary with time of day, season, and period in the sunspot cycle. Government forecasts of radio propagation conditions are published regularly for most amateur bands, and by studying them an amateur can enhance his efforts for DX operation.

## TUNED CIRCUITS

When an alternating voltage is imposed across a capacitor, current will begin to flow as the capacitor charges during the first half

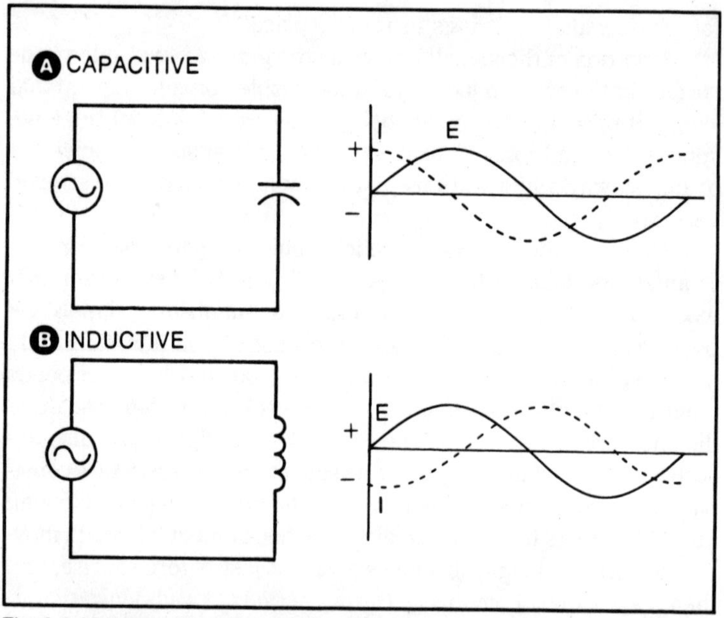

Fig. 9-5. Voltage and current relationships for pure capacitive and pure inductive circuits. With capacitance, current leads voltage by 90 degrees; with inductance, current lags voltage by 90 degrees.

cycle. During the second half cycle the capacitor will discharge back into the source and then charge with opposite polarity. The instant of maximum current is at the start of the half-cycle, not at the instant of maximum voltage. Thus current does flow through the capacitor, but the current sine wave is a quarter cycle ahead of the voltage sine wave—that is, 90° out of phase. Figure 9-5A shows this relationship.

When an alternating voltage is imposed across an inductor current tries to flow, but is opposed by the electromagnetic effect of the inductor. By the time the current flow reaches maximum, the voltage waveform has reached a node. Thus current flows through an inductor, but lags behind the voltage wave by a quarter cycle, as shown in Fig. 9-5B. Again the current is 90° out of phase with the voltage, but in the opposite direction.

Now consider the situation in which an inductor and a capacitor are shunted together in a parallel-tuned circuit. Current flows through both, but one is 180° out of phase with the other. The two currents combine mathematically to result in a current equal to the mathematical difference, with the phase of the larger. We now begin to see that, if the two currents are equal to one another; that is, if the capacitive reactance equals the inductive reactance, something special happens. When the reactances are equal, the inductive current exactly cancels out the capacitive current so that no current flows (see Fig. 9-6). A situation in which voltage is present with little or no current is, by Ohm's Law, a very high resistance or, in the terminology of alternating current, a very high impedance.

Finally, imagine a circuit in which the inductor and the capacitor are in series. If, in a given instance, the current flows first through the capacitor, it will lead the voltage by 90°. Then as it flows through the inductor, the already-leading current will be caused to lag 90° from where it is, putting it into phase with the voltage. When the voltage is in-phase with the current, all that then affects it is the dc resistance of the circuit. Thus a series-tuned circuit offers a very low impedance at resonance.

The above situations have been described for circuits in which there is no dc resistance, only the inductive or capacitive reactance. Such circuits are nonexistent; some dc resistance is always pres-

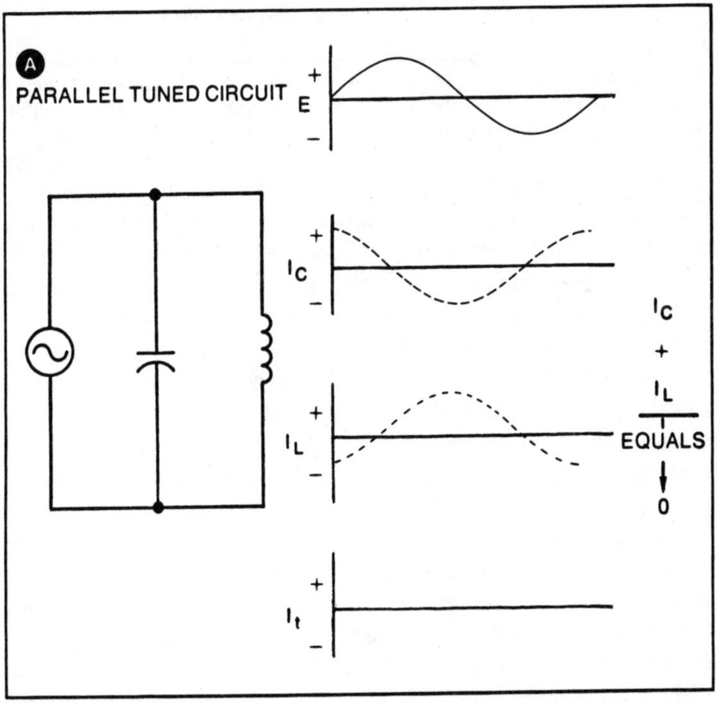


Fig. 9-6. Parallel tuned circuit. Equal and opposite currents are shown flowing through the coil and capacitor. They cancel each other out so that their sum total is zero current.

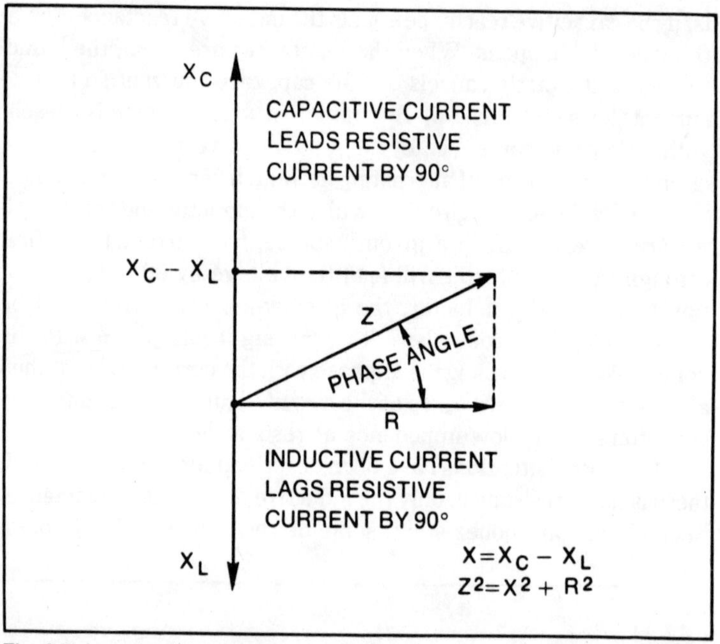


Fig. 9-7. In a circuit having inductance, capacitance, and resistance, the impedance is a composite of the three. The circuit represented by this diagram is nonresonant, as indicated by the net capacitive reactive remaining.

ent. When reactance and resistance combine in an ac circuit they do not add to one another arithmetically, but rather they add as if they were two sides of a right triangle—the impedance of the circuit would be the hypotenuse (see Fig. 9-7). At resonance the reactances combine to cancel one another leaving, in a series tuned circuit, only the dc resistance. In a parallel-tuned circuit the reactive currents add in the same manner as the reactances in a series-tuned circuit. Consequently, a series-tuned circuit has an impedance equal to the dc resistance at resonance, and a parallel-tuned circuit has an equivalent parallel resistance equal to the Q of the circuit multiplied by the inductive reactance.

This figure of merit, Q, is a measure of the relative efficiency of an inductor, and is determined partly by the ratio of inductive reactance to the dc resistance. In a tuned circuit, the relative bandwidth is determined by dividing the resonant frequency by the Q. Figure 9-8 represents this relationship.

## The Antenna as a Tuned Circuit

An antenna is a resonant device and, since the voltage and cur-

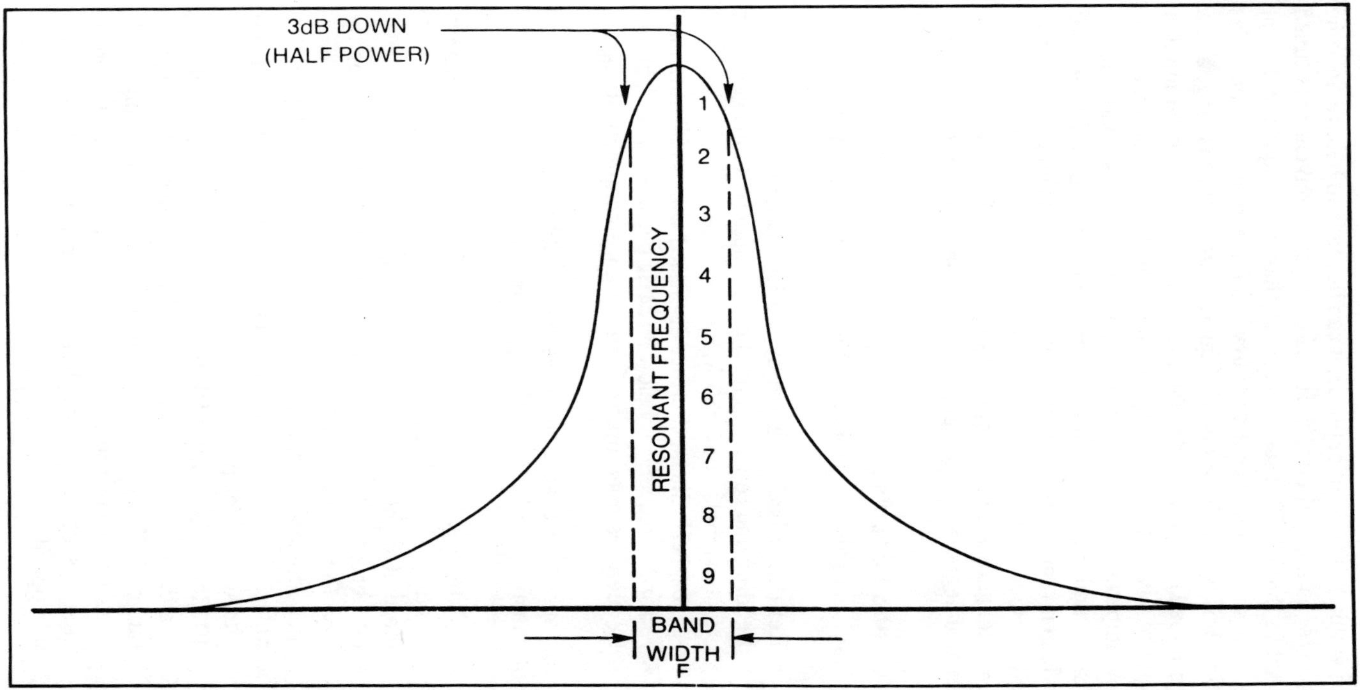




Fig. 9-8. Typical curve of a resonant circuit.

rent within it is 90° out of phase, it can be made to behave, in some respects, as a tuned circuit. It differs from a conventional tuned circuit in that it will resonate at more than one frequency. Fed at the end, an antenna has a very low impedance to all frequencies at which it is an odd number of quarter-wavelengths long, and a high impedance to all frequencies at which it is an even number of quarter-wavelengths long. Fed in the center, an antenna offers a low impedance to all frequencies at which it is an odd number of half-wavelengths long, and a high impedance to all frequencies at which it is an even number of half-wavelengths long. If the feed-point is moved off center, as in a Windom antenna, impedance is proportional to a function of the tangent, based on the number of electrical degrees the feed-point is from the center of the antenna.

## Rules and Formulas

Radio signals are propagated around the world by:

- Reflection back to earth by the ionosphere.
- Bending and reflection by the troposphere. (This generally affects signals above 30 MHz.)
- Atmospheric ducts. (Further discussed with VHF antennas.)
- Aurora. (Frequencies below 30 MHz become "dead"; higher frequencies open up for long-distance communications.)

Ionospheric propagation varies in intensity with the 11-year cycle of sunspots, and affects the amateur bands as follows:

- 160 meters: Reliable daytime band for short-range communications (100 miles or so) day and night. Opens up a little more at night, especially in the winter.
- 80 meters: All right for ranges of 200 miles or so during the day. Better in winter than in summer. Much static on summer afternoons. Nighttime opens up for several thousand miles. Transoceanic operation possible in winter, especially during sunspot peaks.
- 40 meters: Similar to 80, but greter distances possible. Up to a thousand miles or more is possible during the day, with ranges halfway around the world at night. Morning conditions bring out the minimum interference from European broadcast stations. Much less summer daytime static, except in tropical areas.

- 20 meters: One of the best DX bands. Open around the world 24 hours during sunspot peaks. During sunspot lull periods it is dead at night.
- 15 meters: Excellent DX band. More variable than 20, especially during sunspot lull periods, when it "dies" at night, except for occasional sporadic openings. During sunspot peaks, it can be open 24 hours to all parts of the world.
- 10 meters: Generally dead during sunspot lull periods, except for local work. During sunspot peaks, is considered a good daytime DX band, and communications over moderate distances are possible at night.
- 6 meters: We are now getting to the borderline frequencies for ionospheric propagation. DX can be worked during the very peak of the sunspot cycle. Generally offers ranges of one or two thousand miles during sunspot lull periods. Otherwise good for local work.
- 2 meters: Best as a local band. FM operation is very popular. Relatively little ionospheric effects. Interesting tropospheric propagation for the experimenter.
- Above 2 meters: More experimenting done here than in the lower bands. Considerable FM repeater work on 220- and 420-MHz. Tropospheric propagation and atmospheric ducts give interesting results. In the bands above 420 MHz, frequency stability becomes a problem, and communication is generally line-of-sight.

Alternating current and phenomena caused by it are rotating functions. The basic ac generator is a rotating device. Consequently, alternating current devices behave subject to many of the mathematical laws governing circular and angular functions. Each cycle of a wave consists of 360 degrees, and many of the formulas contain functions of pi.

The number pi is approximately:

3.14159 26535 89693 23846 26433 83279 50288
41971 69399 37510 . . .

Don't let the size of the number frighten you. For amateur use, three or four decimal places are plenty.

Radio waves travel at the speed of light. The speed of light in a vacuum is approximately:

186,282.386 miles per second
or
299,793,077 meters per second.

For amateur calculations, you can round these off to 186,000 miles per second and 300 million meters per second.

Except when affected by the atmosphere, radio waves travel in a straight line. That portion of the signal that is closest to the ground is called the ground wave. The portion that radiates at any appreciable angle above ground is called the sky wave. For direct (line-of-sight) communications, the distance to the radio horizon—including "bending" of the signal by the atmosphere, diffraction over the earth's surface, etc.—can be calculated by the formula,

$$D\ (\text{miles}) = 1.415\ \sqrt{H_t} + \sqrt{H_r}$$

$$D\ (\text{kilometers}) = 4.124\ \sqrt{H_t} + \sqrt{H_r}$$

$H_t$ is the height of the transmitting antenna, in feet for the miles formula, in meters for the kilometers formula. $H_r$ is the height of the receiving antenna, in feet for the miles formula, in meters for the kilometers formula.

The vast majority of antennas in use are based on a conductor one half wavelength long. Calculation of a half-wave antenna begins with the formula:

$$\text{Wavelength} = \frac{\text{Velocity}}{\text{Frequency}}$$

Radio waves travel at the speed of light, which is 300,000,000 meters per second. Consequently,

$$\underset{(\text{meters})}{\text{Wavelength}} = \frac{300{,}000{,}000}{\text{Frequency in Hertz}}$$

If we express the frequency in megahertz, we can drop the six least-significant digits from the velocity figure (thus 300). Remember also that we are looking for a half wavelength (300 ÷ 2 = 150). Now we have,

$$\underset{(\text{meters})}{1/2\ \text{Wavelength}} = \frac{150}{\text{Frequency in MHz}}$$

The conversion factor to convert meters to feet is 3.28. Consequently,

$$\text{1/2 wavelength (feet)} = \frac{150 \times 3.28}{\text{Frequency in MHz}} = \frac{492}{\text{F in MHz}}$$

Electricity travels slightly slower in wire than in space. We therefore reduce the velocity factor by about 4.9% (492 × 0.049 = 468), and we obtain,

$$\text{1/2 wave of wire} = \frac{468}{\text{Frequency in MHz}}$$

This is the basic antenna formula, giving a length directly in feet to cut a transmitting (or receiving) antenna that will be resonant, at the desired frequency. There are a couple of other factors affecting the calculation of half-wave antenna. Each type of wire has a slightly different velocity factor, and the feed-point impedance varies with both the height of the antenna and the conductivity of the ground beneath it. To obtain the best possible SWR, along with other improved characteristics, you will have to prune the antenna, that is trim it to exact length required. For this reason, always cut an antenna to be pruned slightly longer than the length obtained from a formula.

Now that we have derived a convenient formula for calculating the length of a half-wave antenna, lets look at some characteristics of other types of antennas.

- The length of an inverted-vee antenna can be calculated from the formula:

$$\text{Length} = \frac{464}{\text{Frequency in MHz}}$$

- A half-wave dipole fed in the center has an impedance of approximately 72 ohms; a half-wave inverted vee fed in the center has an impedance of approximately 50 ohms.
- A quarter-wave antenna fed at the end has an impedance of approximately 34 ohms.
- A quarter-wave ground-plane antenna with horizontal radials has an impedance of approximately 34 ohms.

- A quarter-wave ground-plane antenna with the radials angled down at 45° has an impedance of approximately 50 ohms.
- A half-wave folded dipole has an impedance of approximately 300 ohms.
- A Windom antenna, fed 14% off center has an impedance of approximately 600 ohms.

For efficient transfer of power, the source impedance must match that of the load, and both must be matched by the characteristic impedance of the transmission line. Refer to Table 9-1 for the characteristics of some common types of coaxial lines. If these conditions are not met, an antenna coupler or other matching device is necessary.

The characteristic impedance of coaxial line can be calculated from the formula:

$$Z = 138 \log \frac{D}{d} \quad \text{(for air-insulated line)}$$

Z = characteristic impedance in ohms
D = I.D. of outer conductor
d = O.D. of inner conductor

Diameters can be expressed either in inches or meters, but both must be in the same units.

The characteristic impedance of air-insulated parallel lines can be calculated from the formula:

$$Z = 276 \log \frac{D}{r}$$

Z is the impedance in ohms.
D is the center-to-center distance between the conductors.
r is the radius of the conductor.

**Table 9-1. Coaxial Cable Characteristics.**

| TYPE | Z | MAX WATTS | | | LOSS IN Db/100 | | | CAP/FT pF | VELOCITY FACTOR | DIAMETER |
|---|---|---|---|---|---|---|---|---|---|---|
| | | <30 MHz | 30-150 | >150 MHz | <30 MHz | 30-150 MHz | >150 MHz | | | |
| RG8/U | 50 | 2kW | 1.5kW | 700 | 1 | 2 | 4.2 | 30 | 0.66 | 0.405 |
| RG11/U | 75 | 1.8kW | 1.4kW | 600 | 0.94 | 1.9 | 3.8 | 20.5 | 0.66 | 0.405 |
| RG58/U | 53 | 500 | 300 | 20 | 1.9 | 4.1 | 8 | 28.5 | 0.66 | 0.195 |
| RG59/U | 75 | 800 | 500 | 200 | 1.9 | 3.8 | 7 | 21 | 0.66 | 0.242 |
| TV RG59/U | 72 | 800 | 500 | 200 | 2 | 4 | 7 | 22 | 0.66 | 0.242 |

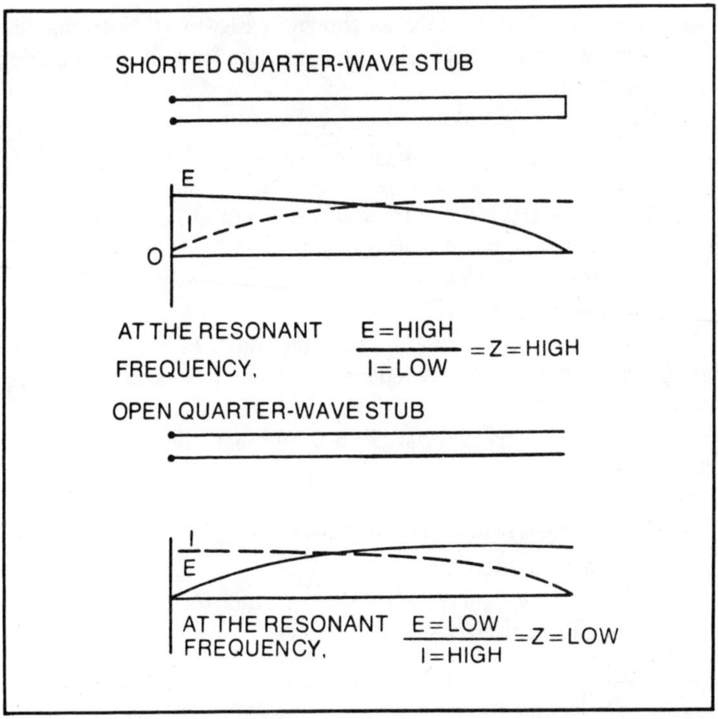


Fig. 9-9. Voltage and current distribution along a quarter-wave transmission line.

As with the previous formula, D and r must be expressed in similar units, either inches or meters.

A length of transmission line cut to a quarter or a half wavelength and used as a tuned circuit is called a stub (Fig. 9-9). Some important characteristics are:

- A quarter-wave stub shorted at the end offers a high impedance to the resonant frequency and shorts out all others.
- A quarter-wave stub open at the end offers a short circuit to the resonant frequency and a high impedance to all other frequencies.
- A half-wave open stub behaves the same as a quarter-wave shorted stub.
- A half-wave shorted stub behaves the same as a quarter-wave open stub.

A quarter-wavelength section of transmission line selected for its impedance can be used as an impedance transforming device by connecting it to the load as shown in Fig. 9-10. Note that the line must be a quarter-wavelength long. The formula for matching the impedance is:

$$Z = \sqrt{Z_s Z_l}$$

Z is the characteristic impedance required for the line.
$Z_s$ is the impedance of the source.
$Z_l$ is the load impedance.

As an example, a 34-ohm antenna and a 75-ohm transmitter output can be matched with a piece of 50-ohm line exactly a quarter-wave long.

Some other characteristics to remember about transmission lines and antennas are:

- A shorted section of line less than a quarter-wavelength long appears inductive.
- An open section of line less than a quarter-wavelength long appears capacitive.

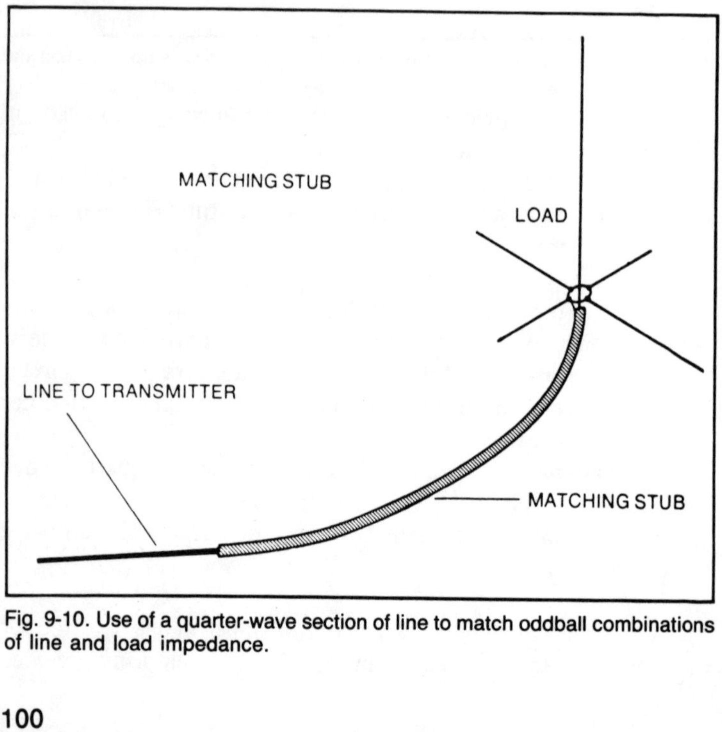


Fig. 9-10. Use of a quarter-wave section of line to match oddball combinations of line and load impedance.

- Any length of line terminated in a resistance equal to its characteristic impedance offers a purely resistive load and reflects no power back toward the source.
- An antenna slightly less than a multiple of quarter wavelengths long appears capacitive.
- An antenna slightly more than a multiple of quarter wavelengths long appears inductive.
- An inductive load is matched by adding capacity.
- A capacitive load is matched by adding inductance.

An inductor offers increasing reactance with increasing frequency. The formula for inductive reactance is:

$$X_L = 2\pi FL$$

$X_L$ is the reactance in ohms.
F is the frequency in megahertz.
L is the inductance in microhenries.

The figure-of-merit, or Q, of an inductor is given by the formula:

$$Q = \frac{X_L}{R}$$

Q is the figure of merit.
$X_L$ is the inductive reactance.
R is the equivalent parallel resistance.

A capacitor offers decreasing reactance with increasing frequency. The formula for capacitive reactance is:

$$X_c = \frac{1}{2\pi FC}$$

$X_c$ is the capacitive reactance in ohms.
F is the frequency in megahertz.
C is the capacity in microfarads (not picofarads).

The inductance of an air-core coil can be calculated by this formula:

$$L = \frac{0.2\ a^2\ n^2}{3a + 9b + 10c}$$

L is the inductance in microhenries.
a is the diameter of the coil in inches.
b is the length of the coil in inches.
c is the depth of the winding (c equals 0 for a single-layer coil).
n is the number of turns.

The capacitance of a capacitor can be calculated from the formula:

$$C = 0.0225K\ \frac{a}{s}\ (N-1)$$

C is in microfarads.
a is the area of one side of a plate. (If two size plates, use smaller.)
s is the space between the plates, or the thickness of the dielectric.
N is the number of plates.
K is the dielectric constant, as given below:

| | |
|---|---|
| Air | 1.0 |
| Bakelite | 4.4 to 5.4 |
| Window glass | 7.6 to 8 |
| Plexiglass | 2.8 |
| Polystyrene | 2.6 |
| Polyethylene | 2.3 |
| Acetate | 3.7 |
| Mica | 5.4 |
| Teflon | 2.0 |

A tuned circuit, consisting of an inductor and a capacitor, is resonant at the frequency at which the inductive and capacitive reactance are equal. Remember that current through an inductor leads the voltage by 90°; current through a capacitor lags the voltage by 90°. With either a perfect inductor or capacitor, current is drawn but no power is consumed.

Resonant frequency is calculated by the formula:

$$F = \frac{1}{2\pi\sqrt{LC}}$$

F is the frequency in megahertz.
L is the inductance in microhenries.
C is the capacity in picofarads.

The following formula can be solved to determine either inductance or capacity:

$$LC = \frac{25330}{F^2} \qquad L = \frac{25330}{F^2\,C} \qquad C = \frac{25330}{F^2\,L}$$

At resonance, a series-tuned circuit offers a very low impedance; a parallel-tuned circuit offers a very high impedance.

The impedance of a nonresonant series circuit is found by the formula:

$$Z = \sqrt{X^2 + R^2}$$

Z is the impedance in ohms.
X is the difference between the inductive and the capacitive reactance.
R is the dc resistance.

To find the impedance of a nonresonant parallel circuit, first combine the reactances with this formula:

$$X = \frac{-X_1\,X_c}{X_1 - X_c}$$

$X_1$ is the inductive reactance.
$X_c$ is the capacitive reactance.

Then combine resistance and reactance in this formula:

$$Z = \frac{RX}{j\sqrt{R^2 + X^2}}$$

The relative strength of a radiated signal is inversely proportional to the square of the distance from the antenna. For example, if the signal strength is known at a given distance from the antenna, at twice the distance, the strength will be one fourth that at the reference point.

Relative signal strength is usually measured in decibels with respect to an established reference set up as 0 dB. The decibel equivalent of a voltage ratio measured across an identical impedance is:

$$dB = 20 \log \frac{E_2}{E_1}$$

The decibel equivalent of a power ratio measured across an identical impedance is:

$$dB = 10 \log \frac{P_2}{P_1}$$

It is much easier to estimate decibels if the relationships in Table 9-2 are remembered.

There is no such thing as a perfect source of electrical energy. Every source, including a radio transmitter, has a certain amount of internal resistance or impedance which consumes some of the power generated. A little application of Ohm's Law with various combinations of load and source resistances will show that the greatest amount of power is delivered when the load resistance is the same as that of the source.

When radio frequency energy is involved, reactive properties in the source and load make the phenomenon far more complex. A certain amount of power is generated, and that which is not consumed by the load is reflected back to the source where it is dissipated in the form of heat. Not only must the source and the load be matched for an efficient power delivery, but also the transmis-

**Table 9-2. dB Equivalents.**

| RATIO | dB POWER | dB VOLTAGE |
|---|---|---|
| 2:1 | 3 | 6 |
| 4:1 | 6 | 12 |
| 10:1 | 10 | 20 |
| 100:1 | 20 | 40 |

sion line characteristics must match or be matched to those of the source and load.

If there is a mismatch, either source-to-load or in transmission line characteristics, the reflected power causes the voltage and current to be unevenly distributed along the line, having maximum and minimum points at intervals coincident with the wavelength of the signal. The ratio between maximum and minimum levels is called the *standing wave ratio*, abbreviated SWR. The SWR can be used as a measurement of the overall behavior of an antenna system. It is measured with a reflected-power meter, or SWR bridge. A ratio less than 2:1 is considered acceptable; 1.5:1 or better is considered pretty good. An SWR of 1.0:1 exists only in theory, and anything approaching that is an indication of excellent performance.

Standing wave ratio is directly proportional to the amount of mismatch in a system. It can be calculated from the formula:

$$\text{SWR} = \frac{R}{r}$$

R is the larger of the two impedances, either source or load.
r is the smaller of the two impedances.

The reflection coefficient can be calculated from the formula:

$$C_r = \frac{R - r}{R + r}$$

r = source impedance
R = load impedance

A negative result of this formula merely indicates a reversal of phase.

The percentage of power being reflected back to the source may be found by squaring the reflection coefficient.

Example: Source = 50 ohms, load = 150 ohms.

$$\text{SWR} = \frac{150}{50} \text{ or } 3{:}1.$$

$$C_r = \frac{50 - 150}{50 + 150} = \frac{-100}{200} = -0.5 \text{ (note phase reversal)}$$

Power reflected back = $(-0.5)^2$ = 0.25, or 25%

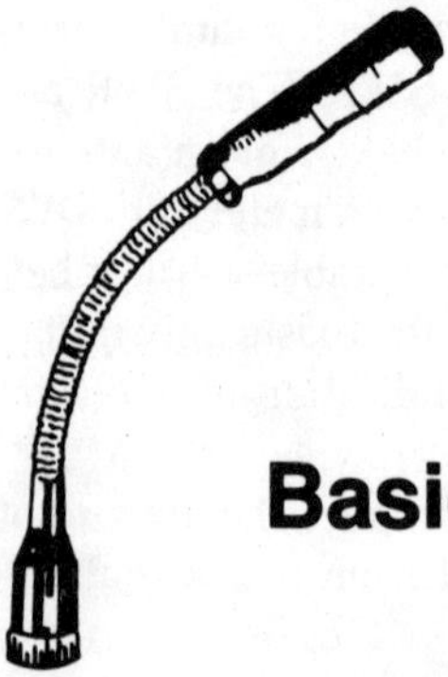

# Chapter 10 Antenna Basics and Examples

YOUR ANTENNA IS THE MOST SENSITIVE PART OF YOUR RAdio communications system; it is the device that interfaces the equipment with the airwaves. Take a moment and think about it. If you were setting up a stereo sound system, would you buy good quality equipment and connect very cheap speakers to it? We can easily draw a parallel between the speakers and the antenna. The antenna is to radio equipment what speakers are to the stereo. Consequently, you must pay careful attention to your antenna.

If an antenna is to work effectively, it must be properly matched to the output circuits of the transmitter, just as speakers must be matched to the stereo output. In addition, the antenna must be resonant (in tune). An antenna can be tuned by properly adjusting its length, or it can be tuned with an inductance and a capacitance. Other circuits can match the output of the transmitter to the antenna's electrical characteristics.

## DETERMINING ANTENNA SIZE

Radio waves travel through space at a constant speed. Therefore, as long as the frequency with which they leave the antenna is constant, there will always be a constant distance between any two successive waves. That distance is dependent on the frequency, and is called the *wavelength*. Radio waves travel through space at 300,000,000 meters per second. If 3,000,000 waves are produced in a second, the wavelength will be 100 meters. We express it this way:

$$\text{Wavelength} = \frac{\text{Velocity (speed)}}{\text{frequency}}$$

When a radio wave strikes an antenna, it produces an electrical voltage. This causes a current to move through the wire. When that current reaches the end of the wire, it has nowhere to go, so it starts back the way it came. The current moves through the wire at a speed just slightly less than the speed of radio waves moving through space. If the antenna length is just right, the current induced by the first wave will have just finished a round trip as the second wave arrives. The precise timing of the waves with the currents in the antenna will cause a buildup in power in the antenna. When this happens, the antenna is said to be *resonant*.

The tuning of an antenna is usually pretty broad, depending on the type of antenna used, so the resonant effect will be realized over a band of frequencies extending on either side of center. Of course, the antenna will work on frequencies outside this band, but not as well.

The resonant frequency of an antenna is a natural function of its length, but it can be changed by adding inductance or capacitance externally. For now, let us stick to natural resonance. This condition occurs when the antenna is one or more quarter wavelengths long, the most prominent resonance being at a half wavelength.

This half wavelength is not half of a wavelength in space, but rather the distance an electrical current at the frequency in question would travel in a piece of wire. The current in the wire travels 2.5% slower than a radio wave in space. Amateurs long ago figured out the following formula for calculating the length of a half-wave antenna:

$$\text{Length in feet} = \frac{468}{\text{frequency in MHz}}$$

## Gain and Decibels

When hams speak of the gain of an antenna, it is not gain in the same sense as that of an amplifier, which adds power to a signal. Antennas cannot do that. They can, however, be designed to collect or radiate more of the signal. The word *gain*, as used here is strictly relative. It refers to the antenna's efficiency as compared to a standard antenna. That standard antenna is a half-wave dipole—an antenna one half wavelength long, fed in the center (Fig. 10-1).

Whether we speak of the gain of an amplifier or the gain of an antenna, we measure it in terms of so many *decibels.* A decibel is one tenth of a bel; a bel is a measure of difference in power.

The word was coined many years ago by telephone linemen to express the amount of loss in telephone wire. It is named in honor of the inventor of the telephone. When radio was invented, the decibel proved to be a handy unit to measure signal power in relation to a known standard. In the case of antennas, this standard is the half-wave dipole.

While there is a simple formula for precisely calculating decibels, the beginner can get along quite well with just a few standard numbers:

- 3 dB represents a 2 to 1 difference in power.
- 6 dB represents a 2 to 1 difference in voltage.
- 6 dB represents a 4 to 1 difference in power.
- 10 dB represents a 10 to 1 difference in power.
- 20 dB represents a 10 to 1 difference in voltage.
- 20 dB represents a 100 to 1 difference in power.

These quantities are represented as positive (+) to indicate an

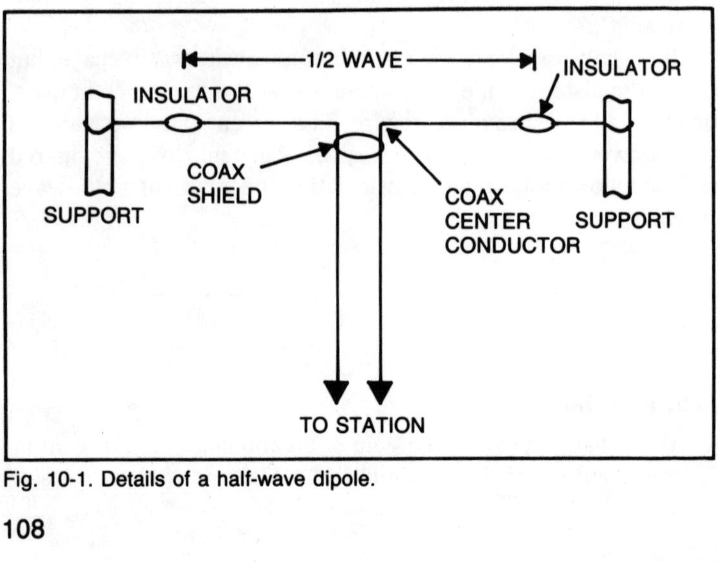


Fig. 10-1. Details of a half-wave dipole.

increase, or negative (−) to indicate a decrease. Knowing these numbers, you can quickly estimate the efficiency of an antenna. If it has a +3 dB gain, it radiates or receives twice as much signal as a half-wave dipole. These expressions are used more at higher frequencies.

## Impedance

Impedance is the opposition a circuit offers to the flow of alternating current. It is different from resistance (its dc counterpart) in that it involves factors that do not affect direct current, and its value is dependent on frequency. Impedance is expressed in terms of ohms the same as resistance.

When one piece of radio equipment is coupled to another, such as a transmitter or receiver being coupled to an antenna, maximum power transfer is realized when the impedance of the two devices is equal. You may already be aware that, in an audio system, one must connect an 8-ohm speaker to an 8-ohm amplifier output. It is the same way with transmitters, receivers, and antennas.

Television receivers are designed to be connected to antennas having an impedance of 300 ohms. Sets made for cable hookups have antenna impedances of 75 ohms.

Most amateur receivers are made to connect to an antenna circuit of 50 ohms impedance. Transmitter impedances are adjustable around a center impedance of 50 ohms. If the impedances of the receiver and the antenna are mismatched by a small amount, it is of no consequence. A large mismatch reduces the efficiency of reception. If the transmitter is mismatched to the antenna, there is much less leeway than with a receiver. Anything more than a slight mismatch produces a phenomenon called *standing waves* which can damage the transmitter circuits and cause interference to nearby TV reception that would otherwise be avoided.

The impedance that an antenna offers to a transmitter or receiver depends on the frequency, the size of the antenna in wavelengths, and the point at which the feedline is connected to the antenna. Since most amateur devices are designed around an antenna system of approximately 50 ohms, it is often necessary to tune the antenna, thereby making it appear to the equipment to be 50 ohms.

Let us examine impedance more closely. If a signal is applied to the end of a wire, current begins to flow as it moves along the wire. When it gets to the end of the wire, it has no place to go,

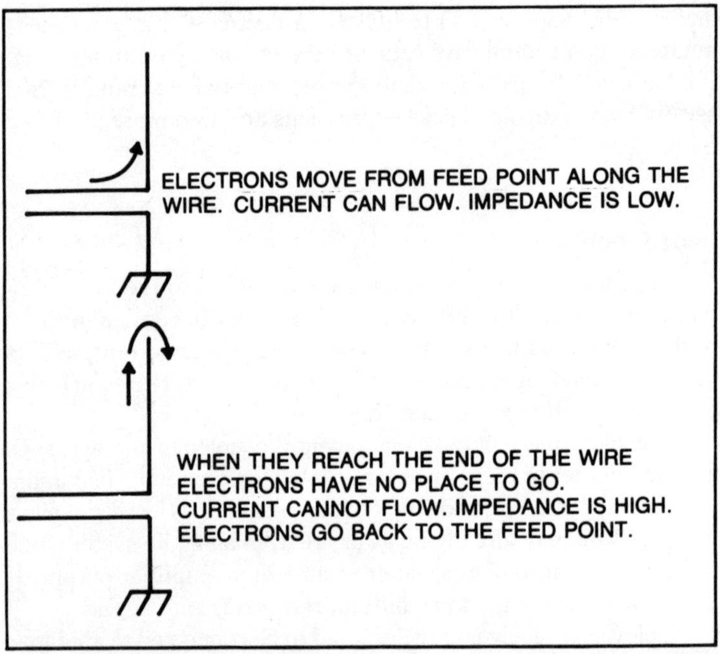

ELECTRONS MOVE FROM FEED POINT ALONG THE WIRE. CURRENT CAN FLOW. IMPEDANCE IS LOW.
WHEN THEY REACH THE END OF THE WIRE ELECTRONS HAVE NO PLACE TO GO.
CURRENT CANNOT FLOW. IMPEDANCE IS HIGH.
ELECTRONS GO BACK TO THE FEED POINT.


Fig. 10-2. Impedance is high at the end of the antenna and low at the feed point, as long as the antenna is a multiple of a quarter wavelength in length.

so it starts back. If the timing is just right and it gets back just as the feed voltage alternates to the other polarity, it can return to the source. If the length of the antenna is exactly right, current flows easily at the feed point. Current flows easily into a low impedance. The size of a wire that will allow this to happen is one fourth of a wavelength. The impedance at the feed point of a quarter-wave antenna is low.

At the far end of the wire, there is a different story. The current has nowhere to go, so it is very low. However, the voltage is still there. High voltage and no current represents a high impedance (Fig. 10-2).

Earlier, you saw a dipole. Normally, a dipole antenna is one half wavelength long. Feeding a dipole at the center, is the same as feeding two quarter-wave antennas set end to end. Just as with the quarter-wave antenna, the half-wave antenna would have a low impedance at its feed point, and a high impedance at either end.

Suppose you tried feeding a half-wave wire at the end. The current fed into the wire would have twice as far to go. By the time it got back to its starting point, it would be opposing the efforts of the transmitter to send the next cycle into the wire. It would still appear as a very high impedance.

From all this, you can see that a half-wave antenna has a high impedance at the ends, and a low impedance in the center. Current is high in the center and zero at the ends; voltage is low in the center and high at the ends (Fig. 10-3).

As a general rule, antennas are fed at a point where the voltage is low and the current quite high. This minimizes problems that might result from arcing, and from the transmission line radiating. When a half-wave antenna is fed at the center, all those conditions are met.

If a center-fed half wave antenna was made of material that was perfect conductor, and if that antenna was someplace in outer space where there was no ground to affect it, the impedance at the feed-point would be zero ohms. In the real world, however, there is no such thing as a perfect conductor, and the proximity of the ground affects antenna impedance. Figure 10-4 shows the effect that height above ground has on antenna impedance.

In addition to having the antenna impedance match that of the equipment, the line that connects the equipment to the antenna must also have characteristics that match that impedance. This fact governs the kind of cable that is used in the system. Coaxial cable is most popular, to the point where you seldom see anything else used.

There are many different kinds of coaxial cable (coax), each having its particular characteristic impedance. This discussion

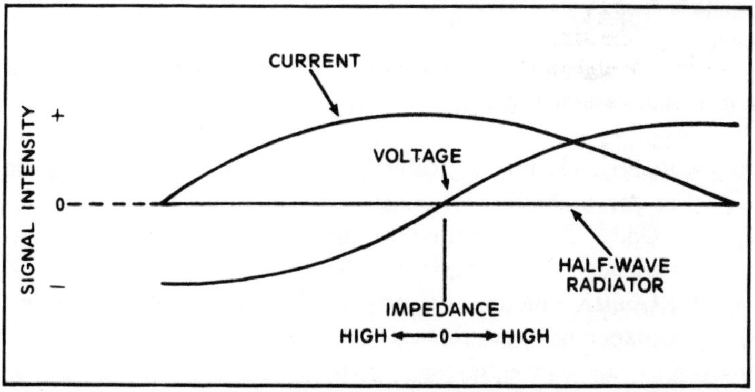


Fig. 10-3. The impedance at the center of a half-wave antenna is theoretically zero. The ideal current and voltage distribution is as shown; voltage is zero at the center, while current is maximum there. At the ends of the antenna, the opposite is true. As you move away from the center, the impedance increases until it is several thousand ohms at the ends of the wire.

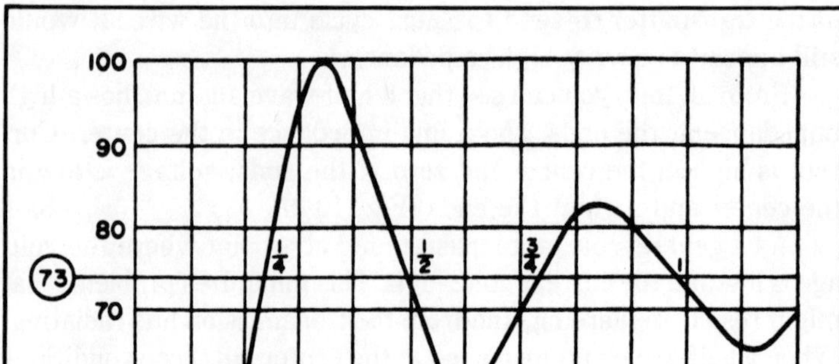

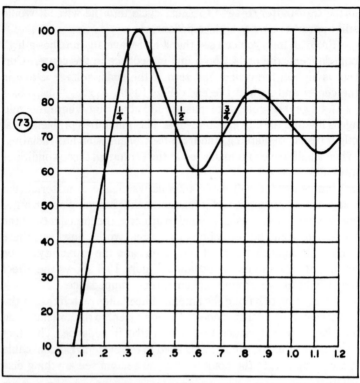


Fig. 10-4. The center impedance of a half-wave antenna illustrated the influence of ground over which the antenna is mounted. The vertical scale is the impedance in ohms, the horizontal scale is antenna height above ground in wavelengths. The impedance fluctuates around a mean of 73 ohms as the antenna is raised.

covers three particular types. These three types are listed in most electronic catalogs in the following manner:

- RG8/U, a heavy-duty cable having a characteristic impedance of approximately 50 ohms.
- RG58A/U, a moderate-duty cable having a characteristic impedance of approximately 75 ohms.
- RG59/U, a moderate-duty cable having a characteristic impedance of approximately 50 ohms.

Most amateur installations use either RG58A/U or RG59/U, except with high power or very high frequencies; while RG8/U, or some of the more expensive cables made for such applications, are used. RG58A/U cable is the better for feeding a half-wave dipole when the antenna is a number of wavelengths above ground. Otherwise, RG59/U is the most popular, general-purpose cable.

Most beginners start out in either the 80 or 40 meter band. Realistic installation conditions generally place the antenna for these bands somewhere between 0.15 to 0.2 wavelengths above the ground. 50 ohm cable is a good match for the range of impedances that Fig. 10-4 shows for these heights. Although the impedance for these conditions varies somewhat from 50 ohms, the equipment is forgiving enough to make that cable quite acceptable.

## STANDING WAVES

So far, we have discussed the need to match the impedance of the transmitter output to that of the antenna, and to match the characteristic impedance of the transmission line to the transmitter and its load (the antenna). If this is not done, you will experience the phenomenon called *standing waves.*

If the wave meets a load that does not match the characteristics of the source and the line—in fact, if there is an impedance mismatch anywhere in the system—a portion of the energy is reflected from the location of the mismatch back along the line to the place whence it came. This energy interferes with energy coming along the line and creates interference patterns within the line called standing waves.

One characteristic of standing waves is that the average signal voltage varies from one point along the line to another. The seriousness of the problem can be expressed by the ratio between the maximum and minimum voltages that would appear if the line were more than a half wavelength long. This ratio is called VSWR, meaning *Voltage Standing Wave Ratio.*

If the entire communications system were absolutely perfect, the VSWR would be 1.0. However, there is no such thing as a perfect system. In the real world, any VSWR of 2.5 or less is acceptable. A well-designed antenna system can have a VSWR immeasurably close to 1.0.

If the VSWR is too high, reflected energy will make its way back into the output circuit of the transmitter and do such nasty little things as burn out parts, and blow transistors. In addition, a high VSWR causes the transmission line to radiate and the energy radiated is usually at frequencies several times the operating frequency. Result? Some angry neighbors when they discover what is happening to their TV sets.

To understand the mechanism of standing waves, look at Fig. 10-5. Since the current at the tip of the antenna is zero, it must be at its maximum value 90 degrees from that point, which in this case is the feed point of the antenna. If the antenna was significantly shorter than a quarter wavelength, the voltage would not be maximum at the end of the antenna, where the current is zero. It would try to increase as the wave were reflected back. If the line was long enough, there would be points where the reflected signal would either be exactly in step with the incoming signal, or exactly out of step. Thus there would be points where the reflected signal were either reinforcing or cancelling the incoming signal.

There are many ways to overcome a mismatch in your antenna system. Various antenna tuning devices are available wherever amateur equipment is sold. They go by various names—match boxes, transmatches, and Antenna Tuners, for example. Most of them work quite well, even some designed for CB use. The only caution is to be sure they will handle the transmitter power. Some CB units will burn up if you feed them more than 50 or 100 watts, especially before they are tuned.

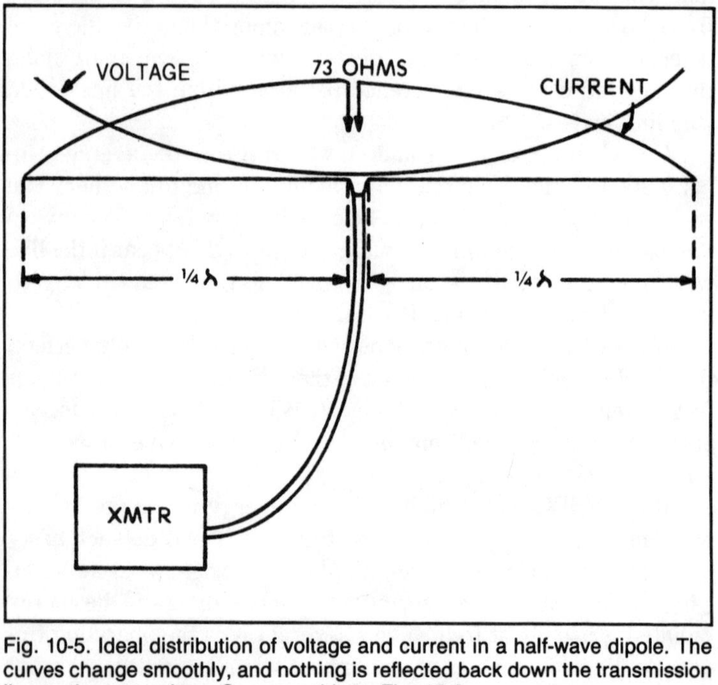


Fig. 10-5. Ideal distribution of voltage and current in a half-wave dipole. The curves change smoothly, and nothing is reflected back down the transmission line to the transmitter. Compare this to Fig. 10-6.

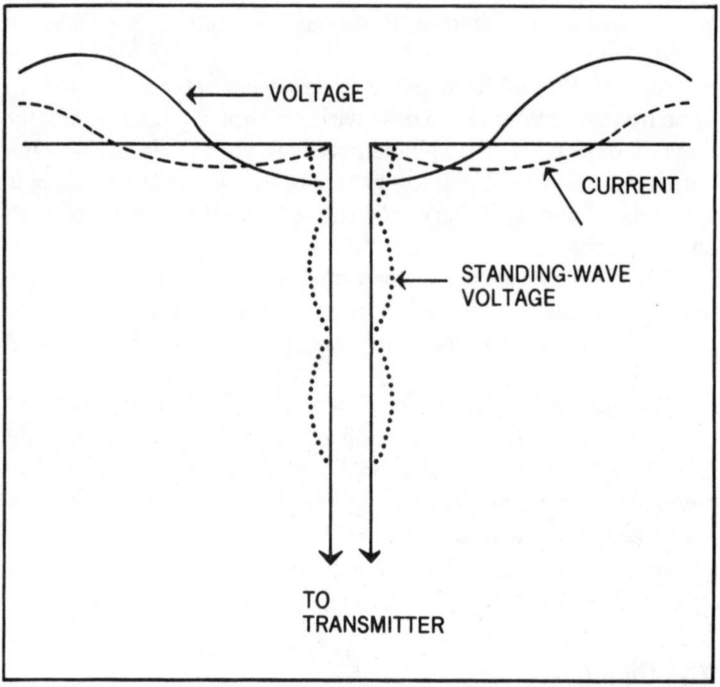


Fig. 10-6. When the antenna is not the proper length for the frequency of the energy feeding it, the curves of voltage and current do not end neatly at the ends of the wire, but see some reactance. This causes some of the energy to be reflected back toward its source. The reflected voltage interacts with the voltage from the source, causing peaks and valleys which are called *standing waves*. The ratio of these peaks to valleys is called *standing wave ratio* (SWR). The two major ill effects of this are that the energy can get into circuits that should not have RF in them; in fact, some modern transmitters are designed to shut down when faced with a high SWR to protect them from burnout.

## PRACTICAL ANTENNAS

Amateurs love to talk about antennas. No matter where you go, you'll find somebody beating his gums to death on the subject. Everybody knows of an antenna that is best for you, and there are nearly as many such antennas as there are hams. You can develop a complete sub-hobby that will last a lifetime just building and trying different kinds of antennas.

While it is generally true that an antenna that works well for transmitting works well for receiving, the reverse is not necessarily so. You can receive signals with just about any piece of wire, whether it is dangling from a tree, strung around the ceiling, or hidden under the rug. Most receivers are sensitive enough that you'll hear plenty of signals. However, transmitting is a different story.

Even if it's well designed, a transmitting antenna can induce radio frequency voltage in house wiring, telephone lines, and other metallic objects. If that voltage gets high enough, it can become not only a nuisance, but a safety hazard as well. If the antenna is seriously mismatched, harmonic energy can raise merry hob with anything from TV sets to computers.

The random wire is the worst offender when it comes to problems like these. It is downright unpredictable. An antenna tuner can be helpful with such an antenna, provided you have a good grounding system.

Ground connections are seldom very efficient, unless you live near a salt-water marsh. Therefore, it's best for beginners to avoid antenna systems that depend on a good ground connection. The beginner can try building his or her own antenna, or there are many excellent kits on the market.

In the following pages, we will discuss some of the popular types of antennas.

## The Dipole

The dipole is one of the easiest antennas to start with. It is simply a half wavelength of wire, supported in a horizontal position, and fed in the center with a suitable transmission line, usually coax. (See Fig. 10-7.)

You can easily calculate the length of a half-wave antenna using the formula given earlier in this chapter. The frequency is usually a number in the center of the band in which you want to operate. Carefully measure the length of wire you need, add about six inches, then cut in the exact center. Place a small insulator in the center as shown in Fig. 10-7, and then connect one conductor of the feedline to each half. Support the antenna as high as possible, letting the feedline drop away as far as practical at right angles to the antenna.

A dipole has an impedance in the neighborhood of 72 ohms, depending on its height. It is generally fed with 75-ohm feedline.

## The Multi-Band Dipole

The dipole is limited to the band for which it was designed, except in unusual circumstances. (A dipole cut for 40 meters also works on 15 meters, due to the relationship of the frequencies.) However, many amateurs work on more than one band. How, then, can an antenna be used on multiple bands? First, one can put tuned wavetraps in the antenna wire to make it appear to end at a different place for different frequencies (Fig. 10-8).

Another way is to use several different dipoles and connect them together. This produces a resonant antenna for each band used, and the antennas not being used present such a high impedance that the transmitter doesn't even notice them (Fig. 10-9).

### The Inverted V

The inverted V is a version of the dipole that has become very popular among amateurs, especially those with limited space. It is a dipole, supported in the center with its ends sloping to the ground (Fig. 10-10). It works best if the angle at the apex is be-

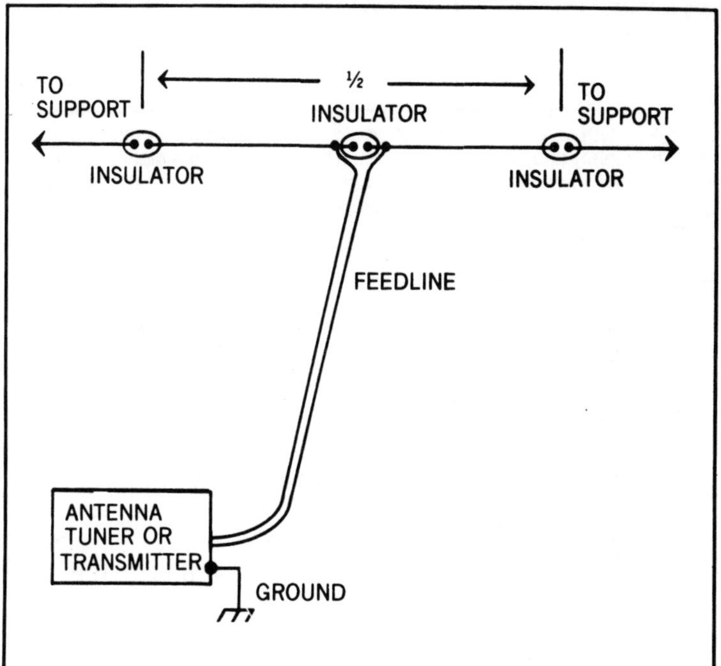


Fig. 10-7. A simple diode antenna is one of the easiest all-around antennas to use, and does not require an extensive ground system; a ground rod connected to your rig with some heavy wire suffices for protection from electrical shock and static buildup.

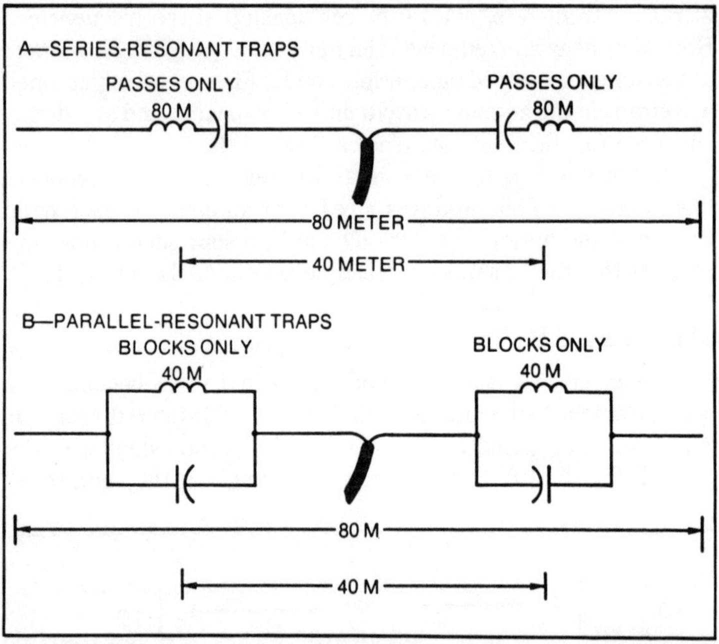


Fig. 10-8. With series traps, the tuned circuits are resonated to pass only the lower frequency (long wire) thus forming a termination to the higher frequency. With parallel traps, the tuned circuits are resonated to block the higher frequency (short wire). Remember, series resonance passes, parallel resonance blocks.

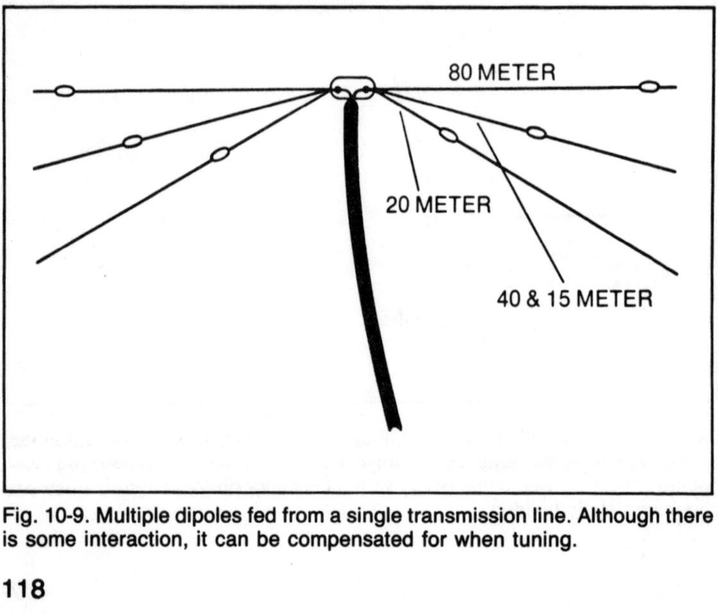


Fig. 10-9. Multiple dipoles fed from a single transmission line. Although there is some interaction, it can be compensated for when tuning.

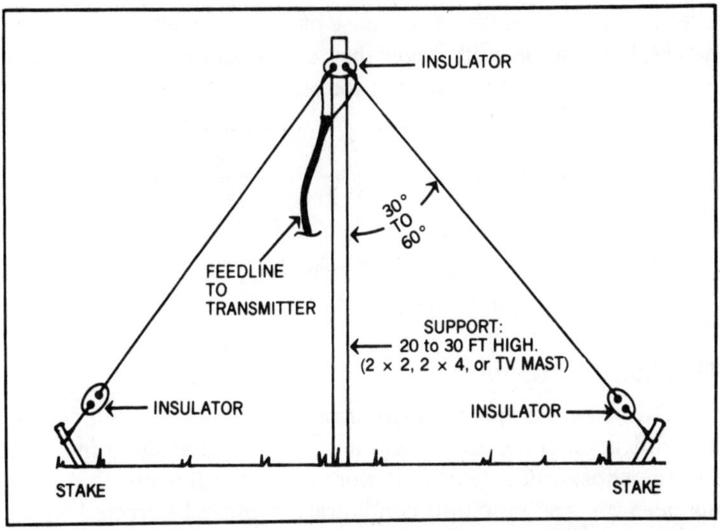


Fig. 10-10. The dipole antenna can be hung from a central support if you don't have two trees or other tall objects for the ends. When built this way, the antenna is called an *inverted V*; it works almost as well as the original dipole.

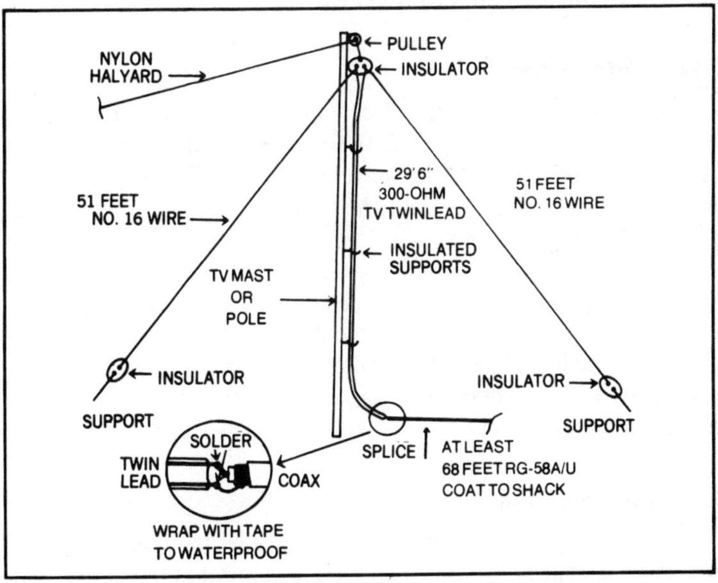


Fig. 10-11. Another version of the dipole is the G5RV. The difference is that the G5RV works on more than one band. Note the precise dimensions given for the wire and the feedline; this is the secret of being able to work on several amateur bands. If you use metal TV masts for support, keep the twinlead 4 to 6 inches from the metal.

tween 90 and 120 degrees. Because of the angle between the two halves, it has to be a bit longer than a straight dipole. The length is calculated with the following formula:

$$\text{Length in feet} = \frac{464}{\text{Frequency in MHz.}}$$

The inverted V has an impedance of approximately 50 ohms, and is fed with 50-ohm coax.

### The G5RV Antenna

This is an example of amateur radio at its best. This antenna, a modification of the dipole, was designed by an amateur in Great Britain whose call is G5RV. It works on several bands, but does not need traps or a multiple configuration. It can be erected either as a dipole or in the inverted V form shown in Fig. 10-11. The dimensions are critical, both the length of the wire and of the feedline. The type of feedline is also very important. Many users around the world swear by its effectiveness.

### The Folded Dipole

This is a quick and dirty receiving antenna that can be made from 300 ohm TV twinlead. With this scheme, the same material is used for both the antenna and the feedline. The physical construction and configuration of the folded dipole is shown in Fig. 10-12. The antenna itself is a half wave section of twinlead. At the exact center of the line, one of the two conductors is cut, as shown in the illustration. The twinlead feedline is connected at the cut.

The folded dipole is an excellent antenna, and easy to build. Because of its impedance, it is recommended as a receiving antenna. However, with a proper antenna tuner, it can work quite well as a transmitting antenna.

### Loaded Antennas

You will often see antennas that have a large lump in the middle. This is especially true with mobile vertical antennas. To the CB operator, such a thing would be called a booster. To the ham, it is a loading coil, or resonator. It consists simply of a coil of wire which has a carefully calculated inductance. The inductance makes the antenna conductor behave as if it were much longer. This ena-

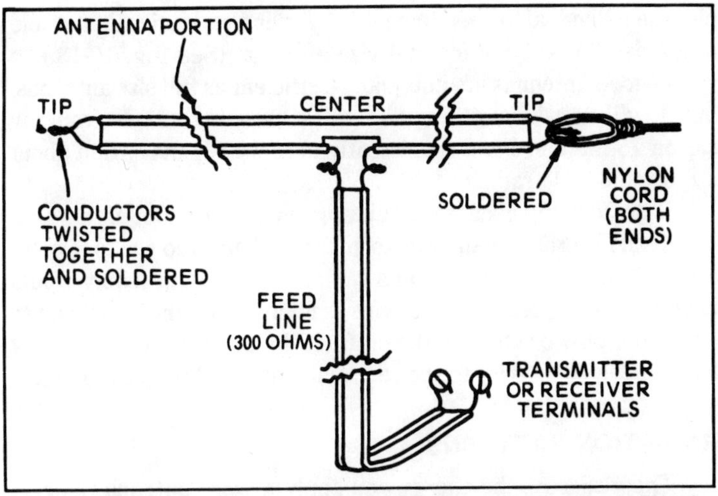


Fig. 10-12. The folded dipole, so called because it is made of common TV twin-lead. The ends are connected together as shown. The feedline is connected at the center, where one dipole wire is broken and spliced to the feedline.

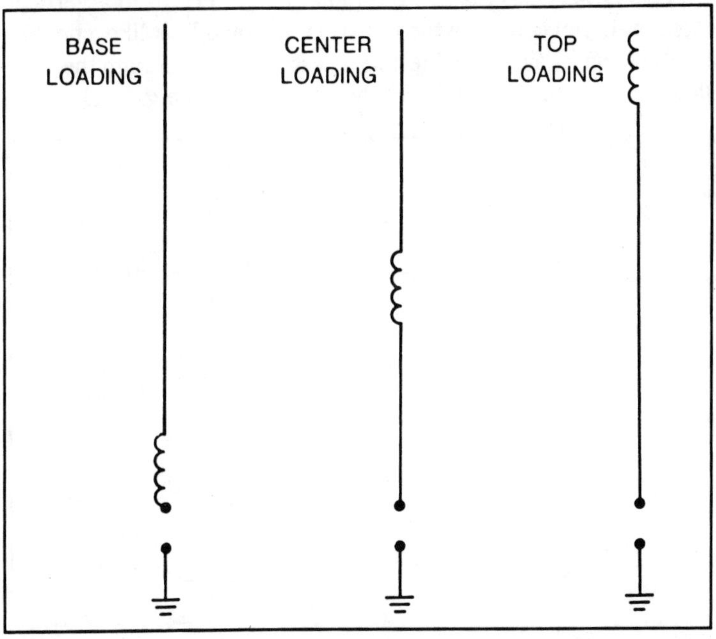


Fig. 10-13. Three methods of loading a mobile whip. Only base and center loading are commonly used; top loading, because of its weight and wind resistance, is rarely seen. Base loading works well for antennas mounted on the roof of the vehicle; center loading performs well with antennas mounted to a fender or bumper.

bles the individual to operate on much lower frequencies than would otherwise be possible for that size antenna. (See Fig. 10-13.)

Loaded antennas are not quite as efficient as full size antennas, but they'll do when nothing else will. Operating mobile from my car on 75 meters, I can consistently chat with hams within about a thousand mile radius.

Besides being used for mobile rigs in an automobile, this kind of antenna is probably the way to go for the ham who wants to operate lowband on the water. On a sailing craft, you probably would have no trouble resonating a wire reaching from the masthead to either the bow or stern of the craft. If the mast is metal, it might lend a little directivity to the radiation pattern, but not too much.

## RADIATION PATTERNS

Theoretically, the energy you apply to your antenna radiates uniformly in almost all directions. In the real world, it is different, due to the proximity of objects on the ground.

It is easiest to visualize the radiation pattern from a vertical antenna. If you could see the radiation, it would look like a doughnut, lying on its side with the antenna sticking up through the middle (Fig. 10-14). In actual use, the nearness of the ground would

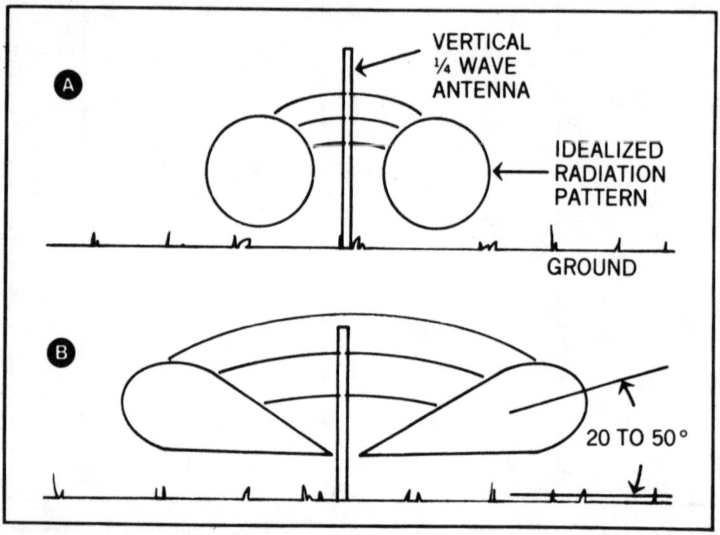


Fig. 10-14. The radiation pattern of a vertical 1/4-wave antenna should look like a doughnut, as shown in A. However, the ground causes some of the energy to be reflected, which distorts the pattern and tilts the maximum radiation upward somewhat, as at B.

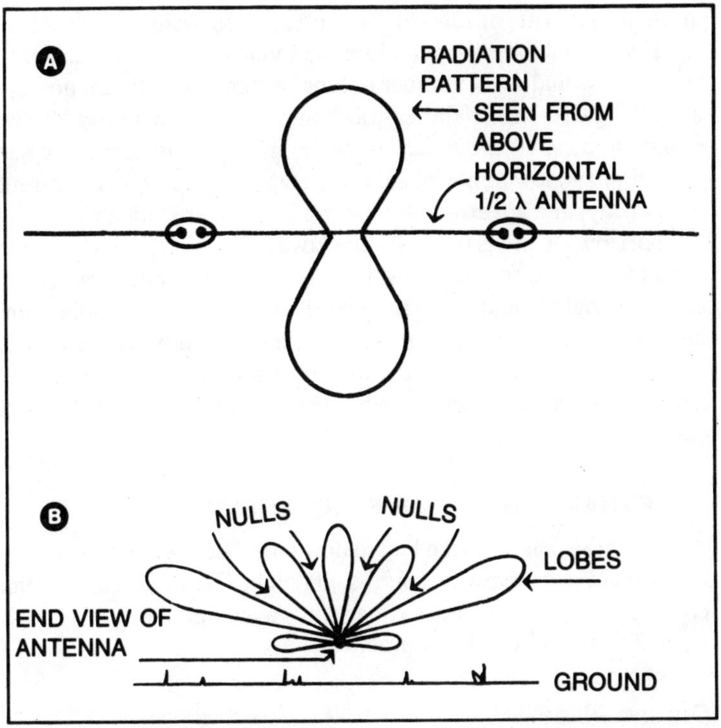


Fig. 10-15. A horizontal dipole antenna in free space ideally has a figure-eight, or hourglass, pattern, as shown by A. Again, the ground below the antenna gets into the act, and reflections cause the pattern to have peaks and nulls, as in B. These peaks shift in position, depending upon how high the antenna is above ground. Picking the right height allows you to get the main lobe at an angle that favors some distances while ignoring others. Most popular antenna theory books have charts that provide this information.

distort the pattern, and the radiation, instead of going straight out, would be at an angle of anywhere from 20 to 50 degrees upward.

It becomes even more complicated with a horizontal antenna. If it were floating in outer space, you would see the familiar doughnut pattern. However, energy reflected from the ground causes a complex pattern which, if viewed from the end of the antenna, would reveal several lobes radiating upward at different angles. Viewed from overhead, two lobes of energy would be seen radiating out at right angles from the antenna, with little or no energy radiating from the ends. See Fig. 10-15A.

Note that most of the energy is radiated at right angles from the wire. (At least, it would seem so if you looked down on it from above.) That is very important when you set up your station. Try to keep that pattern toward the stations you want.

If you could see the radiation, and looked at the antenna endwise, you would see a number of lobes extending outward and upward (Fig. 10-15B). The upward angle of the majority of the radiation is called the *radiation angle.* In the illustration, you see both high and low angle lobes. Each comes into play at a different time of day, and governs whether you could hear stations just over the horizon, or thousands of miles away. High angle radiation is good for stations within a few hundred miles; low angle radiation, under the right conditions, can bring them in from far off while completely missing stations that high angle radiation would bring in. That is why, on the DX bands, we sometimes hear stations clearly from the other coast, while being unaware of stations in the next county.

## ANTENNAS FOR HIGHER FREQUENCIES

The antennas described up to now have been primarily applicable to relatively long wavelengths. On higher frequencies, antenna construction is simplified because the wavelength is short enough to make vertical radiators practical.

### Ground Plane

One of the most popular VHF antennas in existence is the *ground plane,* a quarter-wave, vertical radiator mounted on an artificial ground consisting of three or four horizontally mounted quarter-wave spokes. While this may be a bit tricky to construct for frequencies below 50 MHz or so, it is no problem at all for higher frequencies.

Figure 10-16 is a diagram of the basic ground plane antenna. Note that, while the radials are connected electrically to each other, the vertical member is insulated from them.

Another interesting point about the ground plane: The radials are each a true quarter wavelength long, but the vertical member is about 4.8% shorter. This is because of the capacitive effect of the simulated ground at the feed point.

There are at least two disadvantages in using a ground plane. First, the input impedance is about 35 ohms, making it difficult to find coax that will match it. Secondly, the angle of radiation is a bit high—about 30 degrees. This wastes a lot of usable signal since these frequencies seldom are affected by ionospheric reflection.

Both of these disadvantages can be eliminated by lowering the

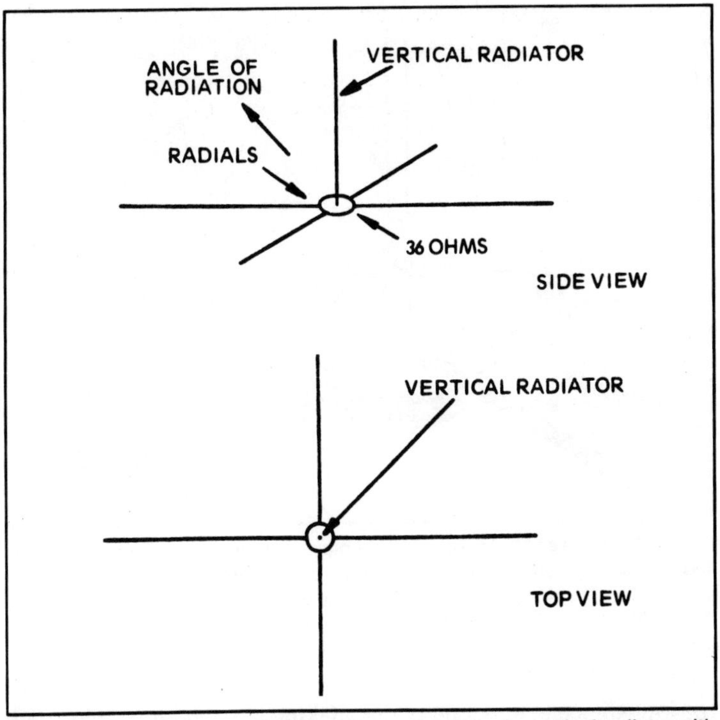


Fig. 10-16. The basic ground plane antenna consists of a vertical radiator with radial wires to simulate actual ground. The radials are all connected together, but are insulated from the vertical element.

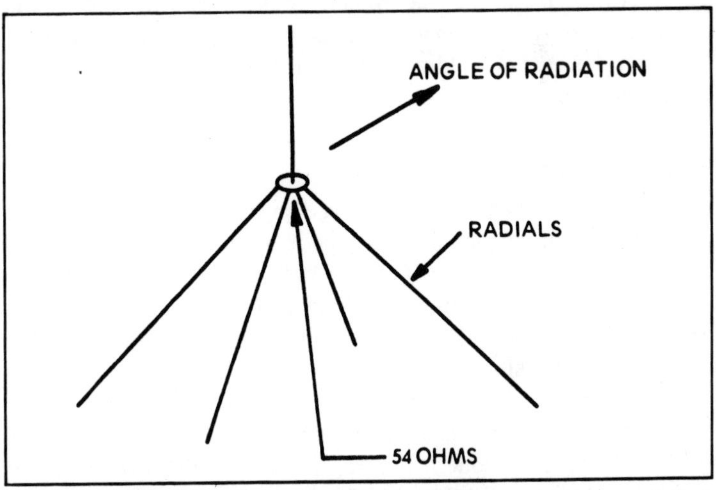


Fig. 10-17. By allowing the radials to droop, the impedance of the ground plane antenna can be made to match 50-ohm coaxial cable.

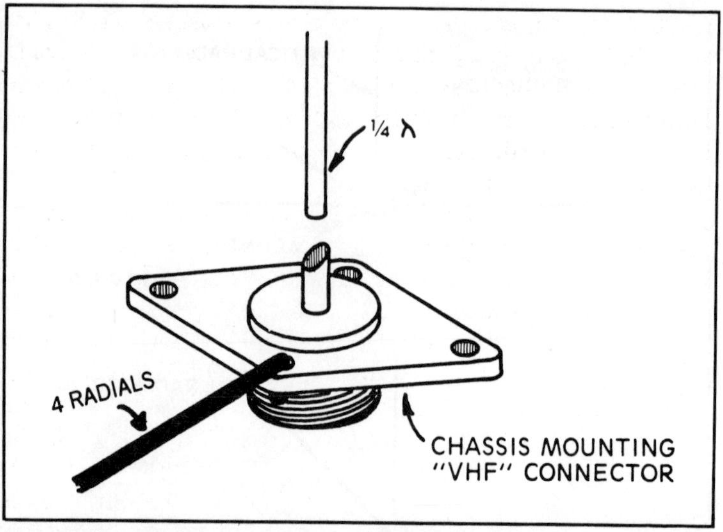


Fig. 10-18. It is easy to make a ground plane antenna for VHF. Simply solder four wires to the shell of a coaxial connector, and another wire to the center conductor. The fitting will mate with the one on the end of your coaxial feed line, and the antenna can be placed high in the air for good coverage.

angle of the radials to about 45 degrees (see Fig. 10-17). This raises the impedance to about 54 ohms, making it a good match for such coax lines as RG8A/U cable. It also lowers the radiation angle to about 25 degrees.

A ground plane for the two meter band can be quickly and cheaply made. All it takes is a chassis-mounting SO-239 connector and a few lengths of welding rod or hard-drawn copper wire. The construction details are shown in Fig. 10-18.

## Coaxial Dipole

The vertical coaxial ship antenna enjoys a great deal of popularity at very high frequencies because it offers an omnidirectional pattern at a very low angle of radiation. Since it is a dipole, its feed-point impedance is 73 ohms, and when a 72-ohm coaxial cable such as RG59A/U is used as a transmission line, a very low standing wave ratio is possible.

If it is made for 2 meters of higher, the coaxial dipole can be made entirely from the coaxial cable that is used to feed it. However, unless you've done it before, it can be difficult to bend half a meter of the coaxial cable shield back on itself, so it may be easier to use a quarter wave section of copper pipe as shown in Fig.

10-20. As you can see, the commercial version uses a quarter-wave whip. Some hams just bend a quarter-wave section of the coaxial cable shielding back over the plastic outer jacket of the cable. This leaves a quarter wave of center conductor, supported by the inner insulation, exposed. The result is the same as the commercial version shown in the illustration.

If you make your own, your best bet is to use solid conductor coaxial cable. The solid inner conductor actually becomes the whip, and the outer conductor becomes the support mast. To prepare the coaxial cable, the outer conductor should be cut about one quarter wavelength back from the cable. This outer metal piece should be removed, leaving the center conductor standing free. A sleeve of 3/4- or 1-inch tubing is then prepared as shown in Fig. 10-21. Prepare the end of the sleeve by making a series of saw cuts as shown. Slip the sleeve over the end of the dipole, bending in the ends as shown, and connect it electrically to the outer jacket of the cable. At the bottom end of the sleeve, make a bushing of plastic tape. This bushing will keep the two separated. You can clamp the aluminum hardline directly to a support below the skirt, without needing insulators.

The coaxial antenna is one of the most versatile and easily con-

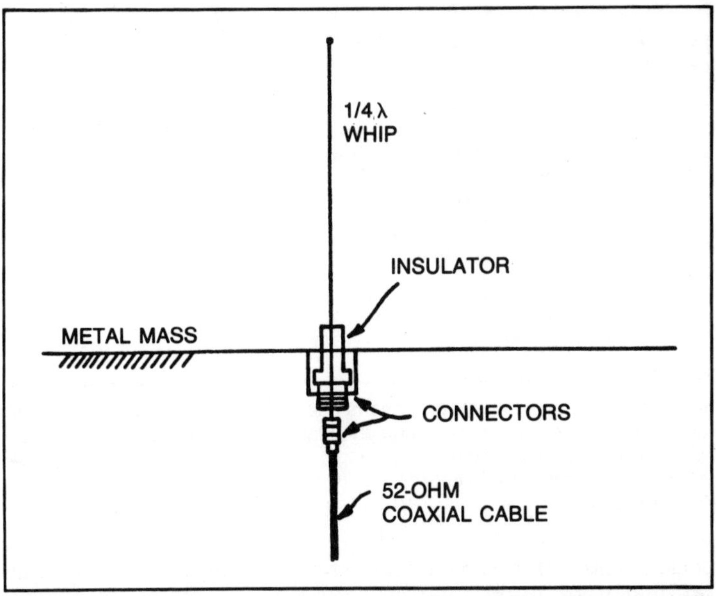


Fig. 10-19. The quarter-wave whip antenna seen on automobiles is actually a ground plane antenna with plenty of metal around it to serve as ground.

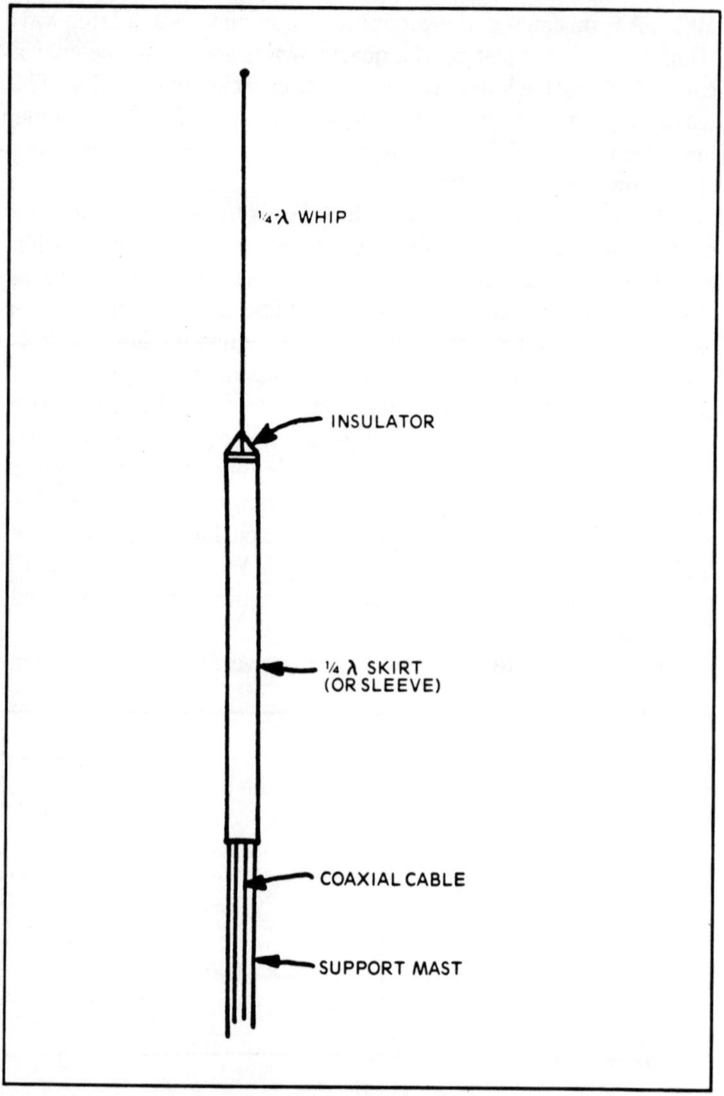


Fig. 10-20. A coaxial dipole antenna is often used for VHF work by amateur and commercial stations. The vertical radiator serves as one half of a dipole, while the coaxial sleeve serves as the other. This type of antenna, when placed high in the air, radiates its energy directly toward the horizon in all directions, instead of at an angle.

structed for VHF use. You can see them at many commercial, military, police, fire, utility, and airport stations, in addition to amateurs who use them at home for their VHF rigs.

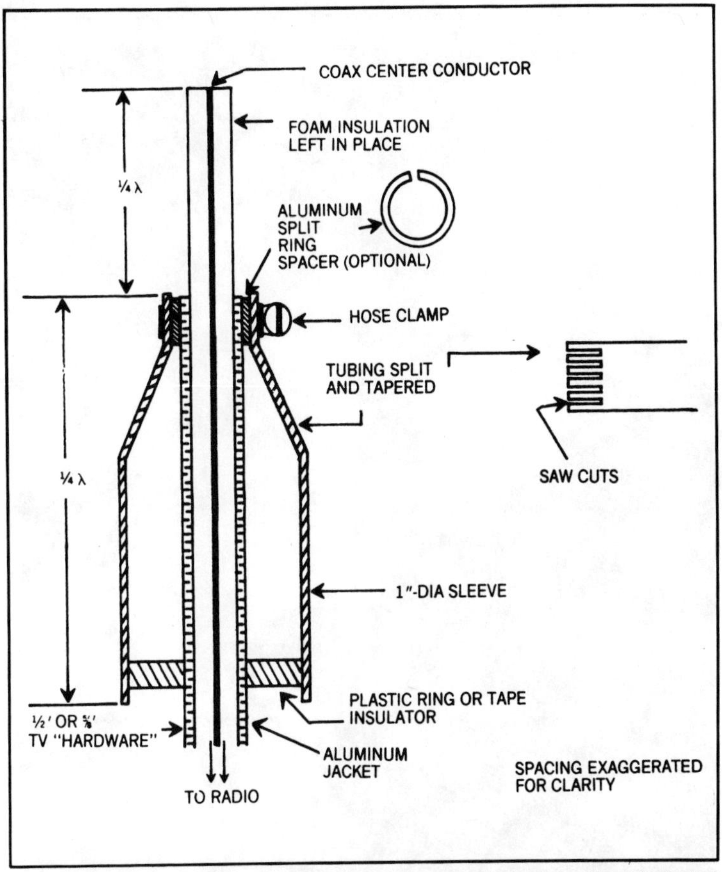


Fig. 10-21. A VHF coaxial antenna can easily be made from aluminum-jacketed cable TV hardline. The top portion of the dipole has the foam insulation still on it. A sleeve of 3/4 or 1 inch aluminum tubing is placed over the hardline to form the skirt. The end of the skirt can be split with several hacksaw cuts to allow it to compress around the hardline. A split ring may be needed to fill the gap if the sizes are too far apart. Layers of plastic tape, or a plastic ring, inside the bottom of the skirt keep it from shorting to the hardline outer jacket.

## The Plywood Tower

Most hams who erect antenna towers use commercially-made ones. However, WA2JHD felt a bit ambitious and built his own from plywood and two-by-fours. The result is shown in Fig. 10-22. It carries antennas for 20, 15, 10, and 6 meters. It was constructed in such a way as to permit easy takedown and setup, and he uses it on Field Day operations as well as at his home station.

Figures 10-23A and 10-23B show a metal tower.

Fig. 10-22. When WA2JHD couldn't afford to buy a tower for his antennas, he built one—out of wood!

Fig. 10-23. Many hands make light work as club members pitch in on the club station antenna (photo courtesy South Towns Amateur Radio Society).

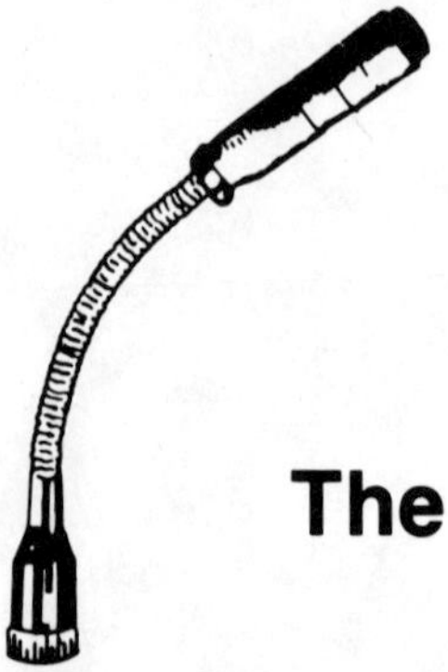

# Chapter 11

# The Mobile Amateur

THERE IS PROBABLY A LARGE NUMBER OF FAMILY MEN WHO have always wanted to become hams but never thought they could afford the time. They often find themselves looking enviously at the fellow who sits at the rig each evening chatting with other hams around the world. But amateur radio, they say, is a time-consuming hobby, and they can't see how they could ever manage to steal enough of this precious commodity from their already busy schedules to make it pay off for them. Their jobs keep them tied up for the better part of the day and, when they get home, they're tired and hungry. After dinner there's a brief period of relaxation with the family, perhaps a little TV, and then it's time to go to sleep again. There just doesn't seem to be any room left for such "isolationist" activities as ham radio.

If this sounds somewhat like the picture of your own life, you ought to consider the prospect of becoming a "mobile only" ham. There are thousands across the country, and the number is growing every week.

## THE PRACTICAL SIDE

The more time you spend in your car, the more practical it is for you to get into ham radio as a mobile amateur. After all, while you are tooling down the highway, doing little but sit and stare out the windshield, you could be thoroughly enjoying one of the more fascinating aspects of the hobby. Long trips can be an awful drag when you have nobody to talk to.

Just in case you're the type that needs to rationalize everything, there are lots of very convincing advantages to mobile operation. Of course, the most obvious benefit is the fact that you will never be completely alone on the road. On long trips it's the disinterested driver that dozes at the wheel. You're less likely to doze off when you're busy chatting with one contact after another. You enter a strange town, and there's always somebody there to direct you to your destination. You break down on the road. CB channel 9 isn't always as well monitored as it should be, but the 2-meter repeaters are. There are few mobile hams who haven't been bailed out of a tight spot at one time or another. I can't count the number of times I've called for and received assistance with any problem ranging from a flat tire to a complete breakdown. A quick "CQ Emergency," or the words "Break, break" spoken into a repeater seldom fail. There isn't a ham in the world who wouldn't make a phone call to the police, or for an ambulance to get help for accident victims.

## THE FUN SIDE

Practical as mobile operation is, it has its fun side as well. For example, transmitter hunting is as traditional to ham radio as hot dogs are to Coney Island. What is a transmitter hunt? Well, it's sort of like playing hide and seek by radio. After all, radio signals are as directive as light beams. During World War II, amateurs had to shut down because of the ease with which enemy planes could home in on their signals.

Leave it to the hams to turn a problem like that around and make a game of it. They play hide and seek with their radios. Whoever is "it" goes and hides, and the others have to home in on his signals and find him. Difficult as that may sound, the average transmitter hunt lasts half an hour or less.

Very few last for hours. Of course, there are exceptions, like the wise guy who rigged up a mobile rig in a kiddie wagon. He went up onto the side of a brush-covered hill, stretched out a dipole in the high grass, and crouched down low. You couldn't see him unless you were on top of him. Lovers were flushed out like quail, as he kept the guys combing the brush on that hillside for hours looking for him.

There was the case of a Pomona, California radio club that had contestants searching for hours around a lake, little realizing that the hidden transmitter was on an island in the middle of the lake. Then there was the time, in Oklahoma City, when a male operator dressed up as a woman and pushed a baby buggy, containing the transmitter, through the park. (There was nothing in the rules that said the transmitter had to remain in one place, but the rules did specify that the rig had to be mobile, and so it was!)

Different clubs, of course, have different rules. A "no-holds-barred" hunt is the exception, rather than the rule. Usually, the transmitter must be within certain distinct boundaries, and hiding on private property is taboo. Some clubs require that the transmitter remain in the car, and that it stay in one place. Then again, there are those that keep rule-making down to the bare minimum necessary to protect the non-participants.

The newcomer to such a scene will give up early, but the experienced hunter can, after making a few U-turns, drive directly to the transmitter location. It's all great fun and, at the end of the hunt, everybody congregates at the favorite fast-food place to rehash the hunt.

There are those that feel this kind of nonsense has its practical application. But in California, a group who call themselves the Happy Flyers have developed the art of locating a downed aircraft to the point where they are often called upon to teach others how it's done. Sometimes they have practice drills; sometimes the real thing. Government regulations require aircraft to have an Emergency Locator Transmitter which, when activated by the impact of a crash, transmit a tone on a distress calling frequency. VHF direction-finding equipment can locate the downed plane in minutes, and rescue efforts can be started. This procedure has saved many lives.

The same procedure also helps track down those characters that get their kicks out of disrupting communications on amateur repeaters. Indeed, those communities whose clubs stage regular "Fox Hunts" have very little trouble with repeater jammers.

Unless it's mounted in the dead center of the roof of a vehicle, and operating at a very short wavelength, a mobile antenna will almost always exhibit directional characteristics. If the receiver has a field-strength meter, finding the direction of the signal can be a snap. Referring to Fig. 11-1, we see that signals are best received along a line containing the greatest mass of metal. Thus, all a hunter must do is make a slow U turn while watching the signal strength meter for the strongest signal. A couple of complete circles com-

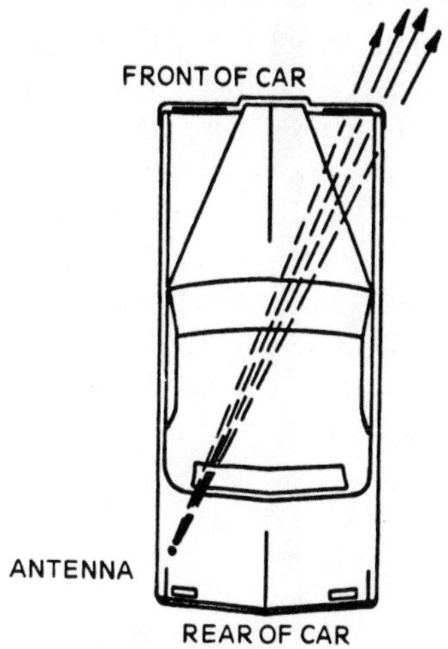


Fig. 11-1. For either transmitting or receiving, the strongest signal exists when the greatest mass of your car lies between the antenna and the station you are communicating with. If you're transmitting, the other station will hear you better in the direction of the arrows. If you're receiving, you can tell the direction the other station is transmitting from by making a couple of 360 degree turns. When the signal sounds best, the other station is in the direction of the arrows.

pletely settle the direction of the signal.

Once the direction has been determined, the hunter simply heads off in that direction. As he gets closer, the signal increases tremendously until it overpowers everything else on the air. Once in a while, the hunter might have to take a second directional reading. Then the hidden transmitter's location is known even more accurately—unless the hider is pulling something sneaky.

## VHF OPERATION

Because of the relatively short wavelength and small antenna size, VHF bands are presently the most popular for mobile operation, and becoming more popular every day. Compact transceivers are available at a moderate cost, and a major portion of the band has been channelized through mutual agreement. As a result of this channelization, every major city has one or more VHF repeaters to enhance mobile operation.

A repeater receives the signal from your mobile unit and

Fig. 11-2. ICOM 2-meter FM handheld transceiver (courtesy of ICOM America, Inc.).

retransmits it, usually 600 kHz higher or lower in frequency. This gives you the advantage of a high-powered station with its antenna positioned on top of a high building or hill. You can then, using only a few watts of power, communicate over hundreds of square miles.

Many repeaters have a built-in phone patch, accessible to members of the owner club who know the correct access tone code.

Through this device, they can dial telephone calls anywhere within the local dialing range of the transmitter site, right from the comfort of their car, or from a hand-held walkie talkie.

Repeaters operate generally in the 2-meter band, although many are becoming available on the 220 and 420 MHz bands. Every major city has a number of 2-meter repeaters capable of covering a radius of up to several hundred miles, depending on the terrain.

## MOBILE INSTALLATION

When you install a mobile station, careful attention must be given to what is called *human engineering*, designing your installation with regard to the physical and psychological needs of the operator. An amateur station, set up at home, has less need for this kind of efficiency, as there is no great consequence if you occasionally have to stretch or reach, or if you get tangled up in a wire. However, you don't want this to happen while you're driving along the interstate.

### Microphone Placement

One of the most often-violated rules of good mobile design is this: Never install a microphone connection in such a position as to allow the cable to come in contact with the steering wheel. The reason is obvious—just try maneuvering a car when there is a mike cable tangled in the wheel. You'll be lucky if you come out of that without an accident. Risky as it is to try maneuvering a car with a mike in your hand, it is even more so when you get tangled in the cord. You could be inviting a visit from the Grim Reaper.

The connector for a hand-held, push-to-talk microphones is best positioned immediately to the right of the driver, and as far back as practical. If your car has bucket seats, the best anchor point is on the console between the seats. This arrangement is easy to install, even if the transceiver is mounted under the dashboard. As shown in Fig. 11-3, you can run an extension mike cable under the console or carpet from the transmitter to the front seat anchor point. Conventional cable clamps secure the cable between the seats.

Use no more cable between the connector and the microphone than is necessary to comfortably reach the operator's face. I recommend coiled cord to keep the slack out of the line and minimize dangerous loops that might get in your way during driving.

Another good approach, but one that has never become very

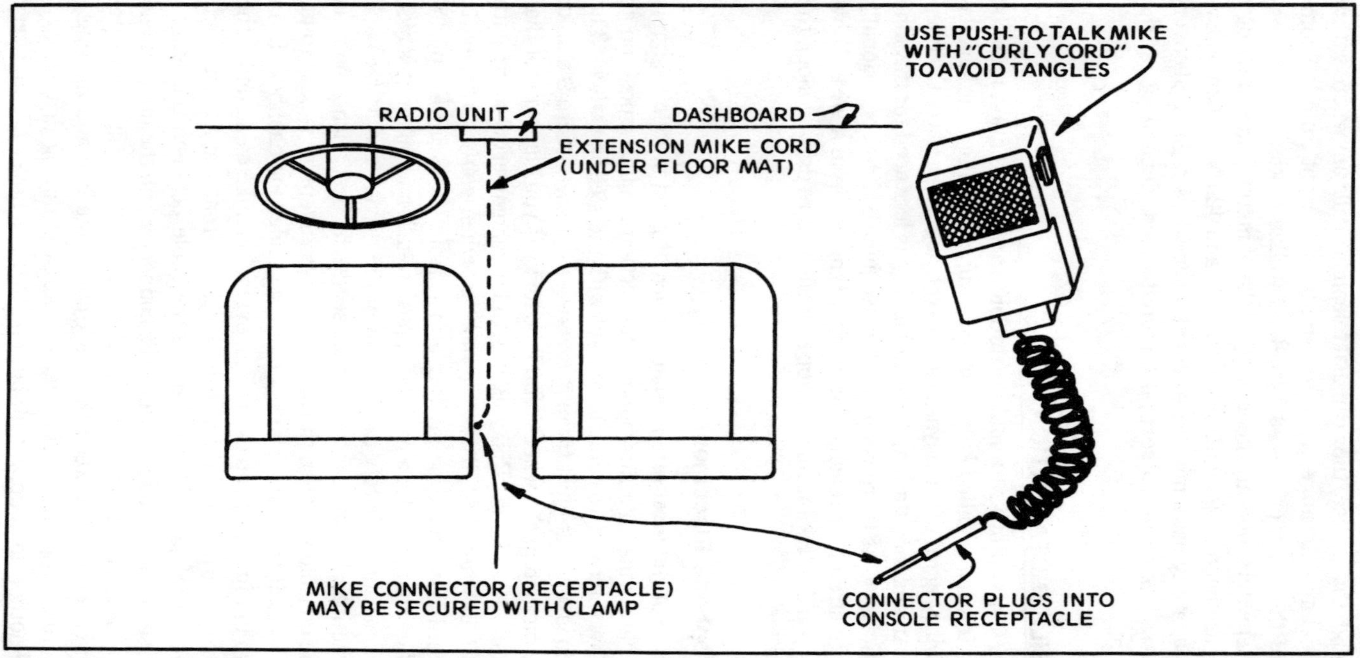
USE PUSH-TO-TALK MIKE
WITH "CURLY CORD"
TO AVOID TANGLES
RADIO UNIT
DASHBOARD
EXTENSION MIKE CORD
(UNDER FLOOR MAT)
MIKE CONNECTOR (RECEPTACLE)
MAY BE SECURED WITH CLAMP
CONNECTOR PLUGS INTO
CONSOLE RECEPTACLE

Fig. 11-3. If the microphone cable terminates on the console between the two front seats, there is little chance of the cord getting tangled with the steering wheel during a turn. Use a standard coiled cable extension, and fasten the receptacle to the console with an ordinary cable clamp.

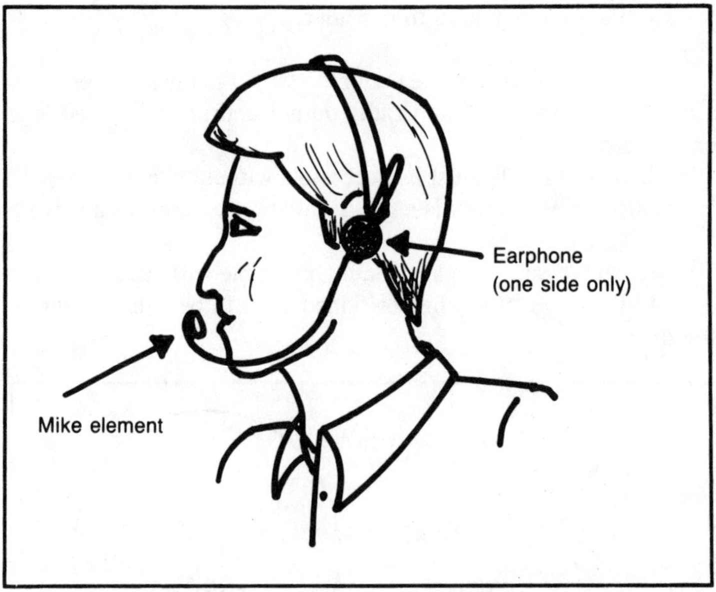


Fig. 11-4. A single-earphone operator's headset is ideal for mobile use, because it keeps both hands free while you're operating. A foot switch or VOX adapter can be used to turn the rig on and off. Headsets are made by companies like HyGain/Telex, and by transceiver manufacturers.

popular, is the telephone operator arrangement, whereby the microphone is incorporated as part of a headset. It is fixed in position several inches from the operator's mouth, as shown in Fig. 11-4. The presence of an earphone offers better hearing in traffic, while one ear is left free to hear those sirens behind you. There is a drawback—no push to talk button—but this can be overcome by installing a foot switch. Some like to use a VOX (voice-operated transmit) system, but I don't recommend it if you often indulge in tirades about other drivers.

## Mounting The Radio

More often than not, the architecture of your car dictates where the rig can be mounted; there aren't any hard and fast rules. The most popular place is under the dash, positioned so that meters, dials, or indicators are clearly visible. Controls should all be within reach without stretching. All this is a tall order in some of today's cars—the best positions are often already occupied. With all the different configurations of rigs and automobiles, it's best to explore the situation and decide for yourself.

Here are some points to consider:

1. The chassis must be out of the way. Family members who are not sympathetic to ham radio do not appreciate injured legs or torn stockings.

2. The controls must be accessible without excessive reaching or stretching. Otherwise, the installation constitutes a driving hazard.

3. Meters and indicators must be visible without stretching.

4. Lights must not be positioned so as to be a distraction to the driver.

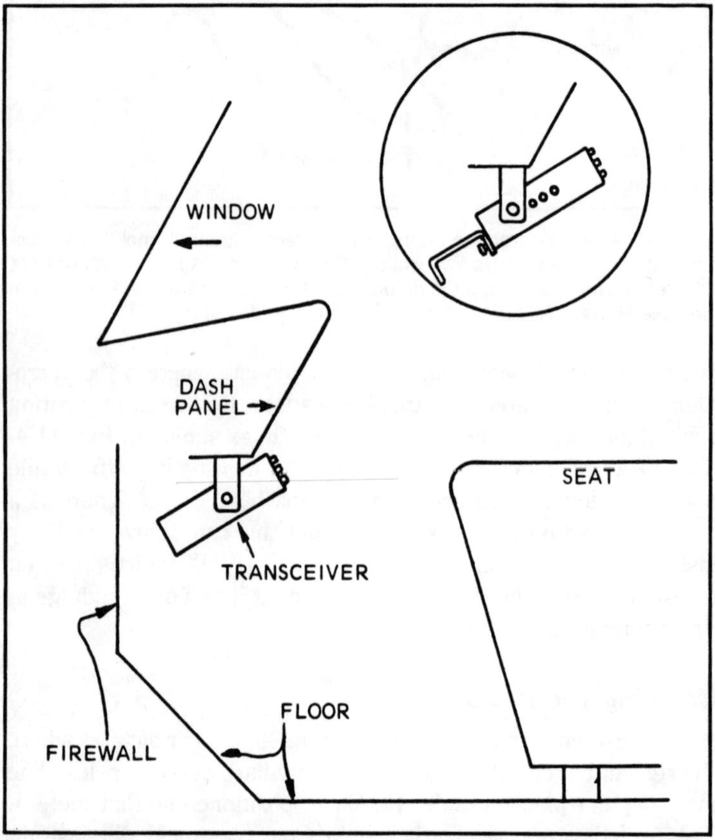


Fig. 11-5. If there doesn't seem to be enough room under the dash for the rig, try mounting it at an angle. You can also slide the rig as far forward as possible in the mounting bracket, as shown in the insert. Some rigs may require an extra brace to eliminate vibration.

5. Microphone cables must be out of the way.

Many of today's transceivers are small enough to easily mount under the dash of the average car. However, dashboards tend to recede toward the front of the car so that radios can't be mounted flat in that position. Angle mounting (Fig. 11-5) is a simple solution to that problem.

Although the transceiver protrudes slightly in front of the dash, secure mounting is possible with an additional bracket in the back of the set. Just be sure there is still plenty of leg room.

## Routing Power Cables

Your transceiver gets its operating power directly from the car battery, and it is best to feed it with as direct a line as possible from the battery. While it seems a good idea to run the power from the ignition switch, the wire to that point is of relatively small gauge, not meant for high-current loads, and is close to other leads that carry an awful lot of electrical noise.

When a transceiver is connected to the ignition switch, the operator experiences difficulty receiving due to excessive noise. When he's transmitting, he notices that the dial lights are noticeably dimmer, and the power output of this rig is not what it should be.

The solution is to install the heaviest possible diameter wire between the battery and the rig. Wire as thick as a quarter inch is neither impractical nor rare. Now, this doesn't mean you can't have the rig interconnected with the ignition switch. It's the way you do it that makes the difference. Use a relay, the heavy-duty type used for starters or horns. You'll need the heavy wire to connect the power to your rig, and you'll need some 16 or 18 gauge automotive hookup wire.

Figure 11-6 shows how it's done. The relay contacts switch the actual power, and the lead from the ignition switch carries relatively little current. You can mount the relay anywhere along the fender wall under the hood, just so long as you can get to it to connect the cables.

Prepare the cables ahead of time by installing solder lugs at every terminal point, as shown in the insert of Fig. 11-6. Connect the ground lead of the heavy wire to the same point as the battery's ground lead. Bring the other end of both leads through a hole in the firewall. Keep the lead lengths as short as possible, while allowing sufficient length to dress the wires out of the way.

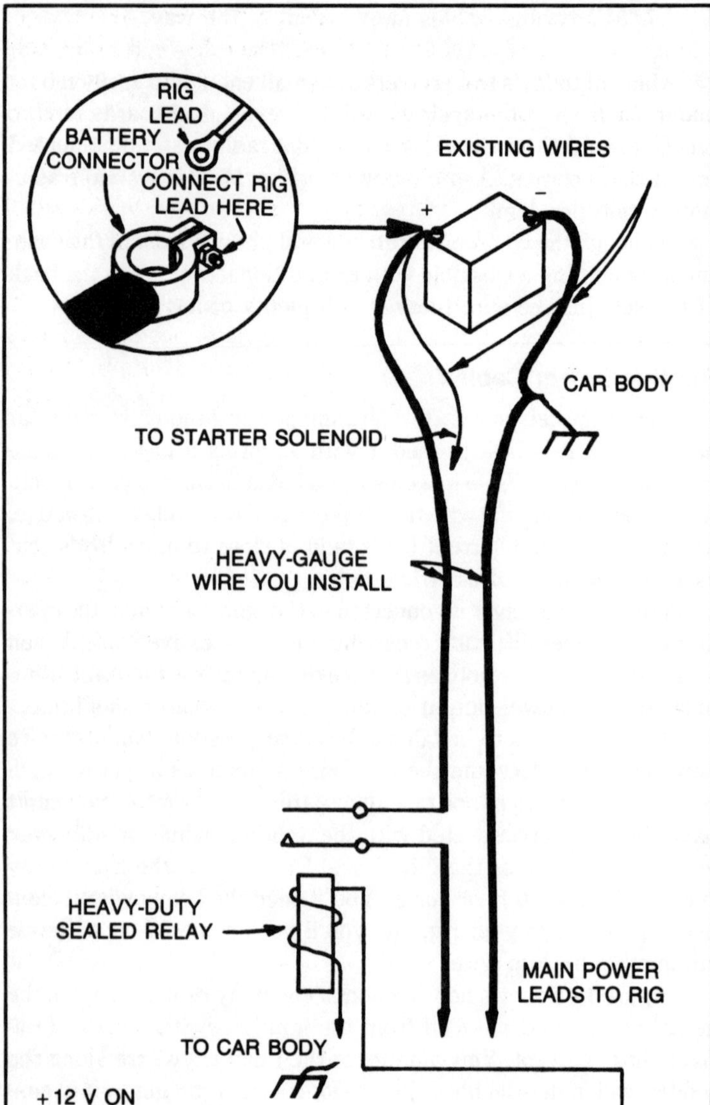


Fig. 11-6. This power-supply scheme will assure full voltage and current for your mobile rig. The heavy relay handles the high current, and the ignition switch merely operates the relay.

Connect the hot battery lead to the positive terminal of either the battery or the starter solenoid. Run the other end to the contacts of the newly-added relay. Run a length of the heavy wire from the other relay contact, through the firewall, to the transceiver.

Connect one of the relay coil contacts to the nearest chassis tie point. Run a wire from the other coil contact to the transceiver on/off switch, using the lighter gauge wire. Run another length of the same wire from the transceiver on/off switch to the battery-voltage terminal of the ignition key.

The solid-state inverter power supplies used with some rigs often have a separate starter wire that must be connected to battery positive in order for the inverter unit to operate. This wire carries very little current and can easily be connected through the ignition switch.

## The Antenna Lead

Since the antenna is often mounted toward the rear of the car, and the transceiver is mounted under the dash, there is always the question of how best to route the transmission line between them. More important than this is which coaxial cable to use. A thinner cable, such as RG-58A/U, offers little problem in routing, since it is small in diameter and fairly flexible. However, thin cable means more loss, and when you get to the distance involved, losses can add up.

Losses in transmission line can be quite significant, especially at wavelengths as short as 2 meters. If you operate on the low bands (10 meters and below) the loss of fifteen feet of so of RG-58A/U will be relatively negligible. At 6 meters, degradation due to line loss reaches the marginal level. On higher frequencies, you must use lower-loss cable such as RG-8/U. This cable is fairly stiff and larger in diameter. It is unwieldy enough to be a royal pain in the neck to install. Now, you don't *have* to use RG-8/U, but there will be a performance tradeoff if you use the smaller cable. The chart in Fig. 11-7 shows the loss that might be expected in an otherwise perfect installation. Remember that a loss of 3 dB means a loss of half your power.

The losses shown in the chart can be greatly increased by heat picked up from the road, or by feeding the line more power than it was designed to handle. (The latter is very unlikely to be encountered in most mobile installations.)

Installing the transmission line is a bother, but if you do it right

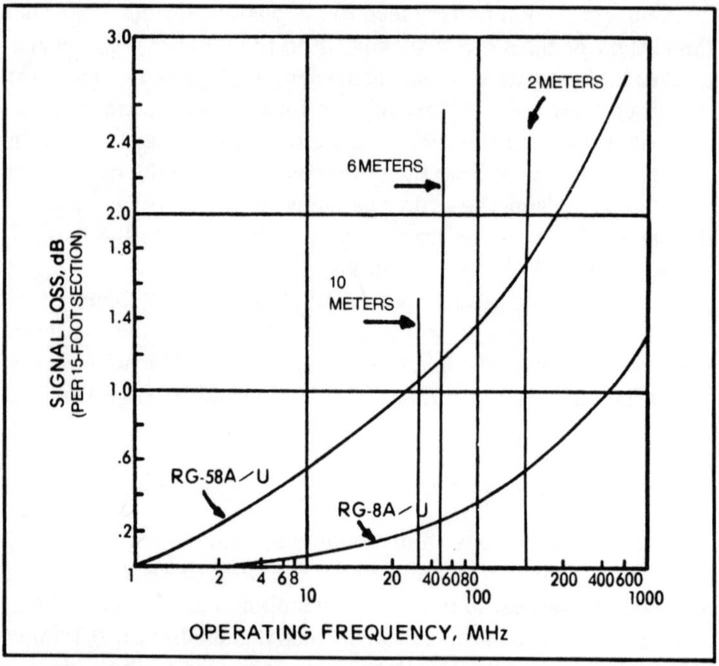


Fig. 11-7. These curves show the power losses that can occur from a 15-foot length of coaxial transmission line. Note that a loss of 3 dB is equivalent to losing half your transmitter power. You'll also lose half the signal you're receiving. Moral: Use the best cable you can get, especially for higher frequencies.

the first time, you'll never regret it. Do it wrong, and you'll be sorry each time you use the rig. The right way (for a trunk-mount antenna) involves feeding the line under the floor mats—back and front—and it involves temporarily pulling out the back seat so you can feed the line into the trunk compartment.

When you prepare the transmission line, install a coaxial connector on one end only. Cut the other end squarely, but don't strip it back or put lugs on it. Connect the cable to the transceiver, then feed the free end under the floor mat from as high up on the firewall as possible.

Temporarily remove the aluminum strip under the door that holds the floor mat in place. Gently peel the mat back so that you can replace it exactly as it was. Feed the transmission line under the mat close to the aluminum strip so that it will not make a bump on the floor. (If you use RG-58A/U cable, the one with greater loss, you can run it under the aluminum strip.)

If the car is a 4-door sedan, use the same procedure under the rear door. Otherwise, you will have to snake the wire under the floor mat to the area beneath the rear seat. At this point it would be wise to remove the rear seat. Once you get the line to the rear seat platform, the rest is easy. There may already be openings giving access to the trunk compartment. Once the line is in place, you can replace the rear seat and the aluminum strips. You will have a nearly invisible installation.

## VHF MOBILE OPERATION

Mobile operation is very different from base-station operation on any band, but there are special considerations to be given the VHF bands that just don't apply to fixed-station operation in these same bands.

On any VHF band, operating range can be greatly increased if you operate from the highest possible elevation. Remember, it's line of sight, and you can see farthest from a hill top. A favorite sport is "hilltopping"—driving to the top of the highest available hill to see how far you can communicate.

VHF mobile operators must consider *antenna polarization*. On these frequencies, the position of the antenna—vertical or horizontal—makes quite a bit of difference. The important thing is that the antennas of both the transmitting and the receiving stations must both be positioned the same way, or there will be a serious loss of signal strength. Mobile antennas are necessarily vertical, even though that results in a higher angle of radiation. Fixed stations that want to work mobile stations should also use vertical antennas.

While there is some SSB and even AM operation in the VHF bands, the overwhelming majority is FM operation. A few years ago, channelized FM stations were crystal controlled, and you had to buy a crystal for each channel you wanted to use. However, nowadays most FM rigs are synthesized, putting any channel or repeater within your reach.

### 6 Meters (AM and SSB)

If you're looking for a real challenge, 6 meters is the mobile band for you. This band, however, is within a natural spectrum for electrical noises. The mobile operator can see that demonstrated vividly. Ignition interference, alternator whine, vibrator hash, distributor-point arcing, leaky lines—they all come in loud and clear on 6!

Indeed, it is a rare case where an amateur can simply install his rig in his car and enjoy operation with no problems. The newcomer to mobile operation will be horrified when he learns that his newly-installed rig can pick up just about everything but legitimate signals. Successful mobile operation usually begins with the tedious task of eliminating noise. Thanks to CB, there are plenty of filters, noise-suppressing spark-plug wire, and other devices that work as nicely on 6 as they do on the citizen's band.

Except for unusual conditions, 6 is a line-of-sight band. Although solar storms and the like occasionally provide spectacular results, most 6 meter operation is limited to 20 miles or less. Operation between two mobile stations is generally 10 miles or less. These broad generalities can be modified upward by such things as well-placed antennas, receiver preamplifiers, very low levels of motor noise, and high transmitter power.

Most of the 6 meter activity is at the low end. This is because TV channel 2 borders the upper end of this band. Most fixed stations stay as far from the TV channel as possible. While mobile operation rarely presents a TVI problem, as the station is seldom in one place very long, mobile operators stay near the low end where there are people to talk to. On the other hand, if you're looking for privacy, and there is no local station operating on channel 2, then the high end of 6 is the place to be.

### 6 Meters FM

Over the past decade or so, FM operation on 6 has become increasingly popular. There are not as many repeaters on this band as on 2 meters, but there are some and the number is increasing. FM operation has the further advantage of being less prone to electrical noise, since most such noise consists of amplitude modulated pulses. A good FM receiver is insensitive to amplitude-modulated noise, but the noise does take its toll by making the receiver less sensitive.

### 2 Meters

This is by far the most popular mobile band. Equipment and antennas are compact, it is well above the noise spectrum, and well separated from any TV channels. Furthermore, there is plenty of action. Although propagation on this band is far more line-of-sight

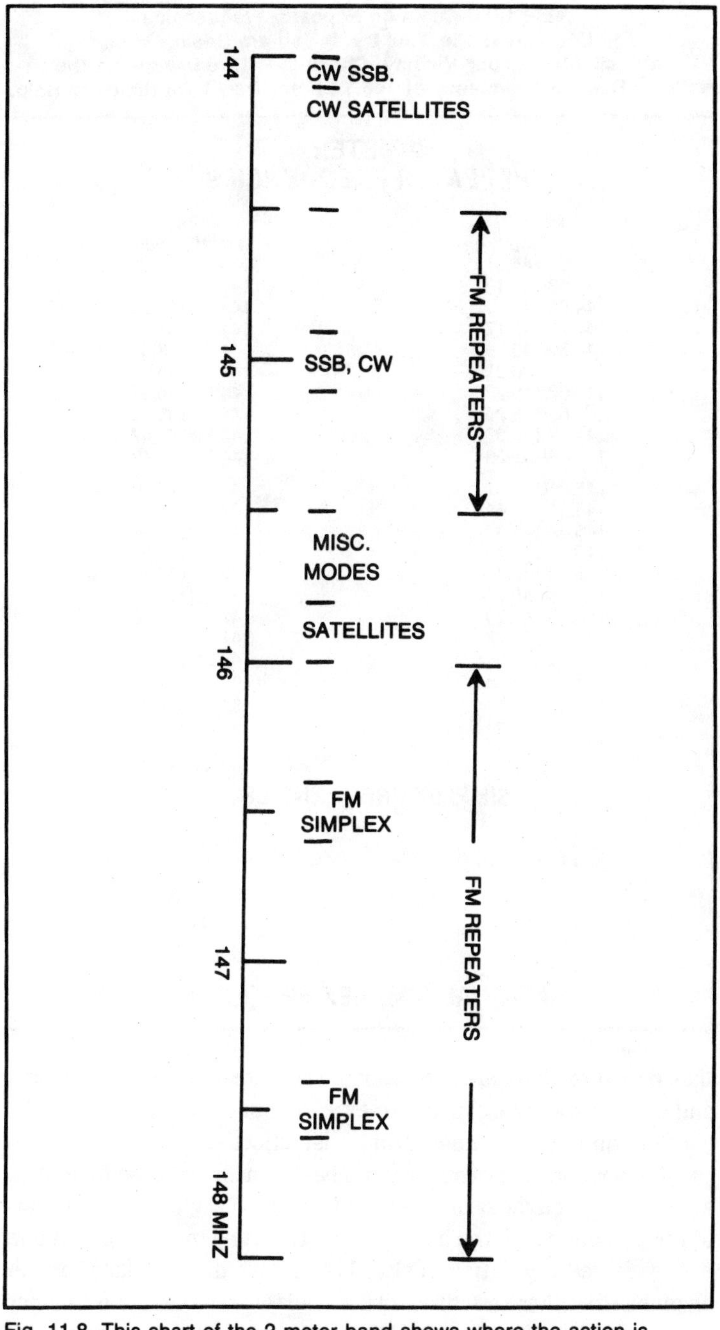


Fig. 11-8. This chart of the 2 meter band shows where the action is.

**Table 11-1. Common Repeater Frequencies in Use Across the Country. If You are Unsure What is Available in Your Vicinity, Check in With Someone on the National Simplex Frequency of 146.52 MHz. They'll Be Happy to Help.**

## 2-METER REPEATER FREQUENCIES

| MHz Input/Output | MHz Input/Output |
|---|---|
| 144.51/145.11 | 144.71/145.31 |
| 144.53/145.13 | 144.73/145.33 |
| 144.55/145.15 | 144.75/145.35 |
| 144.57/145.17 | 144.77/145.37 |
| 144.59/145.19 | 144.79/145.39 |
| 144.61/145.21 | 144.81/145.41 |
| 144.63/145.23 | 144.83/145.43 |
| 144.65/145.25 | 144.85/145.45 |
| 144.67/145.27 | 144.87/145.47 |
| 144.69/145.29 | 144.89/145.49 |
| 146.01/146.61 | 147.00/147.60 |
| 146.04/146.64 | 147.03/147.63 |
| 146.07/146.67 | 147.06/147.66 |
| 146.10/146.70 | 147.09/147.69 |
| 146.13/146.73 | 147.12/147.72 |
| 146.16/146.76 | 147.15/147.75 |
| 146.19/146.79 | 147.18/147.78 |
| 146.22/146.82 | 147.21/147.81 |
| 146.25/146.85 | 147.24/147.84 |
| 146.28/146.88 | 147.27/147.87 |
| 146.31/146.91 | 147.30/147.90 |
| 146.34/146.94 | 147.33/147.93 |
| 146.37/146.97 | 147.36/147.96 |
| | 147.39/147.99 |

## SIMPLEX FREQUENCIES

| | |
|---|---|
| 146.40 | 147.42 |
| 146.43 | 147.45 |
| 146.46 | 147.48 |
| 146.49 | 147.51 |
| 146.52* | 147.54 |
| 146.55 | 145.57 |
| 146.58 | |

***NATIONAL SIMPLEX FREQUENCY**

than 6 meters, there are a great many repeaters in every major city, and even in some rural areas to extend your range. Consequently, it offers an extreme amount of satisfaction.

As mentioned earlier, the band has been channelized by mutual agreement throughout the amateur fraternity. Figure 11-8 is a chart of the present use of the band. Again, this is by mutual agreement, not by forced regulation. Table 11-1 shows the popular repeater channels, and there are directories available that list the frequency and location of thousands of repeaters across the country.

If you are interested in amateur repeater and FM operation, TAB Book No. 1212, *The Practical Handbook of Amateur Radio FM and Repeaters*, contains over 500 pages of information on these subjects.

## HF BAND MOBILE OPERATION

Mobile operation in the 3 to 30 MHz range isn't as popular as 2 meters, but there are a great many followers, and it certainly has its place. A few ambitious souls have tried 160 meters, but antenna and noise problems are considerable.

When you operate mobile on these frequencies, you cannot expect to have a naturally resonant antenna. After all, a quarter-wave whip on 75 meters would have to be about 64 feet long! This problem is met by using a resonating coil in the antenna (Fig. 11-9). Antenna tuning is tricky and, once an antenna is resonated, its bandwidth is very narrow—ten to fifteen kHz on 75 meters is typical.

There is a fair amount of 75 meter mobile activity, in spite of all the obstacles. Operating range is typically in hundreds of miles.

The 40 meter band offers a nice compromise between resonator size, bandwidth, and operating range. It is comparatively free

Fig. 11-9. A homemade antenna for mobile low-band operation. The antenna is set up for 75 meter operation. Whether you build or buy depends on your mechanical skill, your budget, and your aesthetic tastes.

of noise, and a good DX band. Imagine the thrill of working a foreign station from your car!

20 meters is the DX band, although it's not too popular for mobile operation, probably because of less effective local operation. A signal on this band comes in clearly hundreds of miles away, though it may not be heard at all 50 miles away.

10 meters has similar characteristics to the citizens band. While there may be some TVI problems, noise is relatively low and the range is excellent, especially with amateur levels of power. FM operation is growing in popularity on this band, and there are quite a few repeaters in operation. You might want to check out TAB Book No. 1189, *10-Meter FM For The Radio Amateur*.

# Appendix A

# A Study Guide for the Amateur Radio Exams

## QUESTIONS FOR THE ELEMENT 2 AMATEUR RADIO OPERATOR EXAMINATION

The questions used by the examiner in an Element 2 written examination must be taken from the following list. Candidates should NOT TRY TO MEMORIZE THESE QUESTIONS! The examination is a test of the knowledge a person needs in order to operate an amateur radio station properly. It is not a test of a person's ability to answer questions by rote. See PR Bulletin 1035, Study Guide For The Amateur Radio Operator Examinations for more information.

The Element 2 written examination consists of 20 questions. A pass score requires correct answers to at least 15 of the 20 questions. Six or more incorrect answers must be scored as fail.

The examiners may use their discretion as to the format of the answers. They may be single-answer, fill-in-the-blank, multiple-choice, true-false or essay format. See Section 97.27.

SUBELEMENT 2A - Rules and Regulations (7 questions)
One (1) question must be from the following:

2A-1.1 What is the Amateur Radio Service?
2A-2.1 Who is an amateur radio operator?
2A-3.1 What is an amateur radio station?
2A-4.1 What is amateur radiocommunications?
2A-5.1 What is that portion of an amateur radio license that conveys operator privileges called?
2A-6.1 What authority is derived from an amateur radio station license?
2A-7.1 What is a control operator?
2A-7.2 What is the term for the amateur radio operator designated, by the station licensee, to also be

responsible for the emissions from that station?
2A-8.1 What is third-party traffic?
2A-8.2 Who is a third-party in amateur radio communications?

One (1) question must be from the following:
2A-9.1 What are the Novice class operator transmitting frequency privileges in the 80 meter band?
2A-9.2 What are the Novice class operator transmitting frequency privileges in the 40 meter band?
2A-9.3 What are the Novice class operator transmitting frequency privileges in the 15 meter band?
2A-9.4 What are the Novice class operator transmitting frequency privileges in the 10 meter band?
2A-9.5 What, if any, transmitting frequency privileges are authorized to the Novice class operator beside those in the 80, 40, 15 and 10 meter bands?
2A-9.6 In what frequency bands is a Novice class operator authorized to be the control operator of an amateur radio station?
2A-9.7 What does the term frequency band mean?
2A-9.8 What does the term frequency privilege mean?
2A-9.9 In what frequency band is the Novice class operator transmitting frequency privileges 3700-3750 kHz?
2A-9.10 In what frequency band is the Novice class operator transmitting frequency privileges 7100-7150 kHz?

One (1) question must be from the following:
2A-10.1 What is the only emission authorized for use by Novice class operators?
2A-10.2 What does the term A1 emission mean?
2A-10.3 What is the symbol for a transmission of telegraphy by on-off keying?
2A-10.4 What does the term CW mean?
2A-10.5 What, if any, emission privileges are authorized to the Novice class beside A1?
2A-10.6 What is the only telegraphy code a Novice class operator may use?
2A-10.7 Which, if any, telegraphy codes may a Novice class operator use beside the international Morse code?
2A-10.8 What does the term emission mean?
2A-10.9 What is the term for a transmission from a radio station, as used in the FCC Rules?
2A-10.10 What does the term emission privileges mean?

One (1) question must be from the following:
2A-11.1 Under what circumstances, if any, may the control operator cause unidentified radiocommunications or signals to be transmitted from an amateur radio station?
2A-11.2 What is the meaning of the term unidentified radiocommunications or signals?
2A-11.3 What is the term for transmissions from an amateur radio station without the required station identification?
2A-12.1 Under what circumstances, if any, may the control operator of an amateur radio station willfully or maliciously interfere with or cause interference to a radiocommunication or signal?
2A-12.2 What is the meaning of the term maliciously interfere?
2A-12.3 What is the term for transmissions from an amateur

radio station which are intended by the control operator to disrupt other communications in progress?

2A-13.1 Under what circumstances, if any, may the control operator cause false or deceptive signals or communications to be transmitted?

2A-13.2 What is the term for a transmission from an amateur radio station of the word MAYDAY when no actual emergency has occurred?

2A-14.1 Under what circumstances, if any, may an amateur radio station be used to transmit messages for hire?

2A-14.2 Under what circumstances, if any, may the control operator be paid to transmit messages from an amateur radio station?

One (1) question must be from the following:

2A-15.1 What is one of the five principles which express the fundamental purpose for which the Rules for the Amateur Radio Service are designed?

2A-20.1 Call signs of amateur radio stations licensed to Novice class operators are from which call sign group?

2A-20.2 What is the format of a group D call sign?

2A-20.3 What are the call sign prefixes for amateur radio stations licensed by the FCC?

2A-20.4 What determines the number in an amateur radio station call sign?

2A-21.1 With which amateur radio stations may an FCC-licensed amateur radio station communicate?

2A-21.2 With which non-amateur radio stations may an FCC-licensed amateur radio station communicate?

2A-21.3 Under what circumstances may an FCC-licensed amateur radio station communicate with another amateur radio station in a foreign country?

2A-21.4 Under what circumstances (other than RACES operation) may an FCC-licensed amateur radio station communicate with a non-amateur radio station?

2A-21.5 What is the term used in FCC Rules to describe transmitting signals to receiving apparatus while in beacon or radio control operation?

One (1) question must be from the following:

2A-22.1 How often must an amateur radio station be identified?

2A-22.2 How would you identify your amateur radio station communications?

2A-22.3 Do the FCC Rules require an amateur radio station to identify at the beginning of a transmission?

2A-22.4 Do the FCC Rules require an amateur radio station to identify at the end of a QSO?

2A-22.5 What is the FCC Rule for amateur radio station identification?

2A-22.6 What is the least number of times an amateur radio station must transmit its station identification during a 15 minute communication?

2A-22.7 What is the least number of times an amateur radio station must transmit its station identification during a 25 minute communication?

2A-22.8 What is the least number of times an amateur radio station must transmit its station identification during a 35 minute communication?

2A-22.9 What is the longest period of time during a

communication that an amateur radio station does not need to transmit its station identification?

2A-22.10 What is the least number of times an amateur radio station must identify itself during a 5 minute communication?

One (1) question must be from the following:

2A-23.1 What amount of transmitter power may an amateur radio station use?

2A-23.2 What is the maximum transmitter power ever permitted to be used at an amateur radio station transmitting on frequencies available to the Novice class operator?

2A-23.3 What is the amount of transmitter power that an amateur radio station must never exceed when transmitting on 3725 kHz?

2A-23.4 What is the amount of transmitter power that an amateur radio station must never exceed when transmitting on 7125 kHz?

2A-24.1 Should an amateur radio operator receive an Official Notice of Violation from the FCC, how promptly should he/she respond?

2A-24.2 Should an amateur radio operator receive an Official Notice of Violation from the FCC, to whom does he/she respond?

2A-24.3 Should an amateur radio operator receive an Official Notice of Violation from the FCC relating to some violation that may be due to the physical or electrical characteristic of the transmitting apparatus, what information must be included in the response?

2A-25.1 Whom does the FCC hold responsible for the proper operation of an amateur radio station?

2A-25.2 When must an amateur radio station have a control operator?

2A-25.3 Who may be the control operator of an amateur radio station?

SUBELEMENT 2B - Operating Procedures (1 question)

One (1) question must be from the following:

2B-1.1 What does the S in the RST signal report mean?

2B-1.2 What does the R in the RST signal report mean?

2B-1.3 What does the T in the RST signal report mean?

2B-2.1 At what telegraphy speed should a CQ message be transmitted?

2B-3.1 What is the meaning of the term zero beat?

2B-3.2 Why should amateur radio stations in communication with each other zero beat?

2B-4.1 How can on-the-air transmitter tune-up be kept as short as possible?

2B-5.1 What is the difference between the telegraphy abbreviations CQ and QRZ?

2B-5.2 What is the difference between the telegraphy abbreviations K and SK?

2B-5.3 What is the meaning of the telegraphy abbreviations DE, AR, and QRS?

SUBELEMENT 2C - Radio Wave Propagation (1 question)

One (1) question must be from the following:

2C-1.1 What type of propagation uses radio signals refracted back to earth by the ionosphere?

2C-1.2 What is the meaning of the term skip propagation?

2C-1.3 What is the area of weak signals between the ranges of ground waves and the first-hop called?

2C-1.4 What is the meaning of the term skip zone?

2C-1.5 What does the term skip mean?

2C-1.6 What type of radio wave propagation makes it possible for amateur radio station to communicate long distances?

2C-2.1 What type of propagation involves radio signals that travel along the ground?

2C-2.2 What is the meaning of the term ground wave propagation?

2C-2.3 Daytime communication between two stations on 3.725 MHz is probably via what kind of propagation when they are located a few miles apart but separated by a low hill blocking their line-of-sight path?

2C-2.4 When compared to skip propagation, what is the usual effective range of ground wave propagation?

SUBELEMENT 2D - Amateur Radio Practice (3 questions)

One (1) question must be from the following:

2D-1.1 How can an amateur radio station be protected against being operated by unauthorized persons?

2D-2.1 Why should all antenna and rotor cables be grounded when an amateur radio station is not in use?

2D-2.2 How can an antenna system be protected from damage due to a nearby lightning strike?

2D-2.3 How can amateur radio station equipment be protected from damage due to lightning striking the electrical wiring in the building?

2D-3.1 For proper protection from lightning strikes, what pieces of equipment should be grounded in an amateur radio station?

2D-3.2 What is a convenient indoor grounding point for an amateur radio station?

2D-3.3 To protect against electrical shock hazards, the chassis of each equipment in an amateur radio station should be connected to what?

2D-4.1 When working on an antenna mounted on a tower, a person doing the climbing should always wear what type of safety equipment?

2D-4.2 For safety purposes, how high should all portions of a horizontal wire antenna be located?

2D-4.3 What type of safety equipment should a person on the ground wear while assisting another person on an antenna tower?

One (1) question must be from the following:

2D-5.1 What is a likely indication that radio frequency interference to a receiver is caused by front-end overload?

2D-5.2 When radio frequency interference occurs to a receiver regardless of frequency, while an amateur radio station is transmitting, what is likely the problem?

2D-5.3 What type of filter should be installed on a television receiver's tuner input as the first step in preventing overload from an amateur radio station's signal?

2D-5.4 What is meant by receiver overload?

2D-6.1 What is meant by harmonic radiation?

2D-6.2 Why is harmonic radiation by an amateur radio station undesirable?

2D-6.3 A multi-band antenna connected to an improperly tuned transmitter may radiate what type of interference?

2D-6.4 What is the purpose of properly shielding a transmitter?

2D-6.5 When interference is observed on only one or two channels of a TV receiver, while an amateur radio station is transmitting, what is the likely problem?

2D-6.6 What type of filter should be installed on an amateur radio transmitter as the first step in reducing harmonic radiation?

One (1) question must be from the following:

2D-7.1 Why is it important to have the impedance of a transmitter final-amplifier circuit match the impedance of the antenna or feedline?

2D-7.2 What is the term for the measurement of the impedance match between a transmitter final-amplifier circuit and the antenna or feedline?

2D-7.3 What station accessory is used to measure RF power being reflected back down the feedline from the transmitter to the antenna?

2D-7.4 What station accessory is often used to measure voltage standing wave ratio?

2D-7.5 Where should a SWR bridge be connected to indicate the impedance match of a transmitter and an antenna?

2D-7.6 Coaxial feedlines are designed to be operated with what kind of standing wave ratio?

2D-7.7 If the SWR bridge reading is higher at 3700 kHz than at 3750 kHz, what does this indicate about the antenna?

2D-7.8 If the SWR bridge reading is lower at 3700 kHz than at 3750 kHz, what does this indicate about the antenna?

2D-8.1 What kind of SWR meter reading may indicate poor electrical contact between parts of an antenna system?

2D-8.2 High SWR readings measured from a half-wave dipole antenna being fed by coaxial cable can be lowered by doing what to the antenna?

SUBELEMENT 2E - Electrical Principles (3 questions)

One (1) question must be from the following:

2E-1.1 Electrons will flow in a copper wire when its two ends are connected to the poles of what kind of source?

2E-1.2 The pressure in a water pipe is comparable to what force in an electrical circuit?

2E-2.1 What are the two polarities of a voltage?

2E-2.2 What type of current changes direction over and over again in a cyclical manner?

2E-2.3 What is a type of electrical current called that does not periodically reverse direction?

2E-3.1 List at least four good electrical insulating materials.

2E-3.2 List at least three good electrical conductors.

2E-3.3 What is the term for the lowest voltage which will cause a current in an insulator?

2E-4.1 What is the term for a failure in an electrical circuit that causes excessively high current?

2E-4.2 What is the term for an electrical circuit in which there can be no current flow?

One (1) question must be from the following:

2E-5.1 When a voltage is applied to a circuit

causing an electrical current to flow, what is consumed?

2E-6.1 What is the approximate length, in meters, of a radio wave having a frequency of 3.725 MHz?

2E-6.2 What is the relationship between frequency and wavelength?

2E-6.3 What is the approximate length, in meters, of a radio wave having a frequency of 21.120 MHz?

2E-7.1 Which are higher: radio frequencies or audio frequencies?

2E-7.2 Is 3,500,000 Hertz a radio frequency or an audio frequency?

2E-7.3 Radio frequencies are considered to be those above what frequency?

2E-8.1 Are audio frequencies higher or lower than radio frequencies?

2E-8.2 Audio frequencies are considered to be those below what frequency?

2E-8.3 What frequency range is 2500 Hertz: audio or radio?

One (1) question must be from the following:

2E-9.1 What is the unit of electromotive force?

2E-10.1 What is the unit of electrical current?

2E-11.1 What is the unit of electrical power?

2E-12.1 What is Hertz the unit measurement of?

2E-12.2 What is another popular term for Hertz?

2E-13.1 A frequency of 40,000 Hertz is equal to how many kilohertz?

2E-13.2 A current of 20 millionths of an ampere is equal to how many microamperes?

2E-13.3 A current of 2000 milliamperes is equivalent to how many amperes?

2E-13.4 What do the prefixes mega and centi mean?

2E-13.5 What do the prefixes micro and pico mean?

SUBELEMENT 2F - Circuit Components (1 question)

One (1) question must be from the following:

2F-1.1 What is the general relationship between the thickness of a quartz crystal and its fundamental operating frequency?

2F-1.2 What is the schematic symbol for a quartz crystal?

2F-1.3 What chief advantage does a crystal controlled transmitter have over one controlled by a variable frequency oscillator?

2F-2.1 What two internal components of a D'Arsonval meter interact to cause the indicating needle to move when current flows through the meter?

2F-2.2 What does a voltmeter measure?

2F-3.1 Draw the schematic diagram of a triode vacuum tube and label the elements.

2F-3.2 Draw the schematic symbol for a tetrode vacuum tube and label the elements.

2F-3.3 What was one of the earliest uses of a two-element vacuum tube?

2F-4.1 What device should be included in electronic equipment to protect it from damage resulting from a short circuit?

2F-4.2 When an excessive amount of current flows through a fuse, what happens to the fuse? the circuit? the current?

SUBELEMENT 2G - Practical Circuits (1 question)

One (1) question must be from the following:

2G-1.1 Draw a block diagram representing the stages in a simple crystal-controlled telegraphy transmitter.

2G-1.2 What type of transmitter does this block diagram represent?

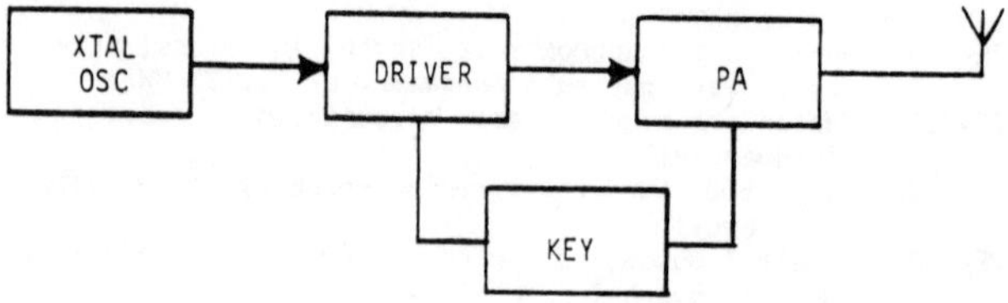


2G-1.3 Draw a block diagram representing the stages in a simple telegraphy transmitter having a variable frequency oscillator.

2G-1.4 What type of transmitter does this block diagram represent?

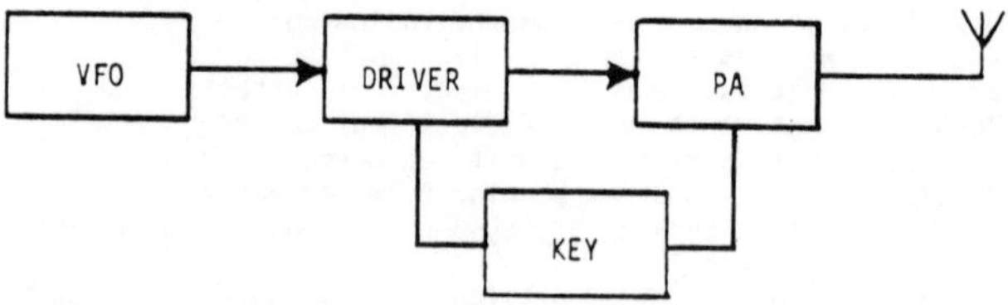


2G-2.1 Draw a block diagram representing the stages in a simple superhetrodyne receiver capable of receiving A1 telegraphy signals.

2G-2.2 What type of device does this block diagram represent?

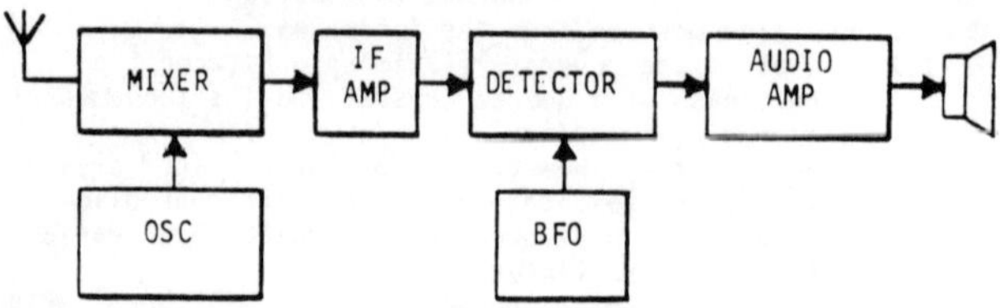


2G-3.1 Draw a block diagram representing how two different antennas and a dummy load can be connected to the same transceiver.

2G-3.2 What is the unlabeled block in this diagram?

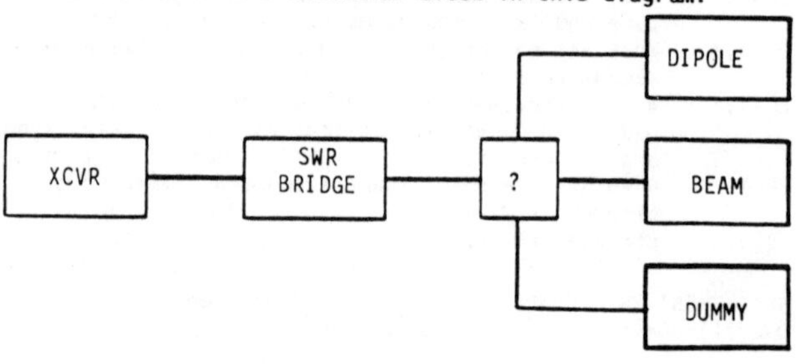

2G-3.3 Draw a block diagram representing an amateur radio station including a transmitter, receiver, telegraph key, TR switch, SWR bridge, antenna tuner and antenna?

2G-3.4 What is the unlabeled block in this diagram?

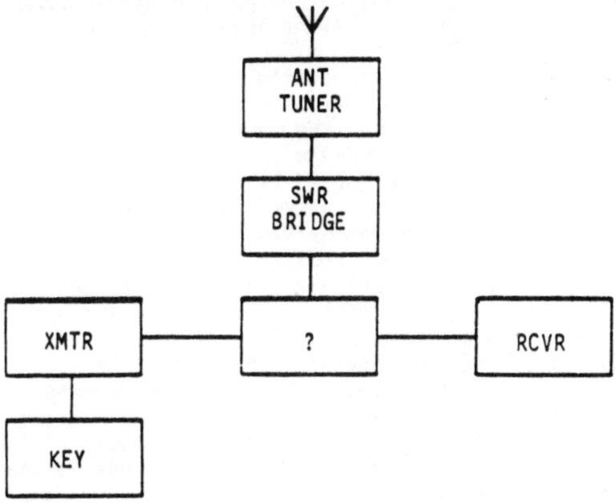


SUBELEMENT 2H - Signals and Emissions (1 question)

One (1) question must be from the following:

2H-1.1 An interrupted carrier wave is considered to be which type of emission?

2H-2.1 What does the term backwave mean?

2H-2.2 What is a possible cause of backwave?

2H-3.1 What does the term key clicks mean?

2H-3.2 How can key clicks be eliminated?

2H-4.1 What does the term chirp mean?

2H-4.2 What can be done to a telegraph transmitter's power supply to avoid chirp?

2H-5.1 What is a common cause of superimposed hum?

2H-6.1 A signal received on 28.160 MHz is the 4th harmonic of what fundamental frequency?

2H-7.1 Spurious emissions from a transmitter may be caused by what problem in the power amplifier stage?

SUBELEMENT 2I - Antennas and Feedlines (2 questions)

One (1) question must be from the following:

2I-1.1 What is the approximate total length in feet of a half-wave dipole antenna cut for 3725 kHz?

2I-1.2 What is the approximate total length in feet of a half-wave dipole antenna cut for 7125 kHz?

2I-1.3 What is the approximate total length in feet of a half-wave dipole antenna cut for 21,125 kHz?

2I-1.4 What is the approximate total length in feet of a half-wave dipole antenna cut for 28,150 kHz?

2I-1.5 How is the approximate total length in feet of a half-wave dipole antenna calculated?

2I-2.1 What is the approximate total length in feet of a quarter-wave vertical antenna adjusted to resonate at 3725 kHz?

2I-2.2 What is the approximate total length in feet of a quarter wave vertical antenna adjusted to resonate at 7125 kHz.

2I-2.3 What is the approximate total length in feet of a

quarter-wave vertical antenna adjusted to resonate at 21,125 kHz?

2I-2.4 What is the approximate total length in feet of a quarter-wave vertical antenna adjusted to resonate at 28,150 kHz?

2I-2.5 When a vertical antenna is lengthened, what happens to its resonant frequency?

One (1) question must be from the following:

2I-3.1 What is a coaxial cable?

2I-3.2 What kind of antenna feedline is constructed of a center conductor encased in insulation, which is then covered by an outer conducting shield and weatherproof jacket?

2I-3.3 What are some advantages in using coaxial cable as an antenna feedline?

2I-3.4 What commonly-available antenna feedline can be buried directly in the ground for some distance without adverse effects?

2I-3.5 When an antenna feedline must be located near grounded metal objects, which commonly-available feedline should be used?

2I-4.1 What is parallel conductor feedline?

2I-4.2 Can an amateur radio station use TV antenna "twin lead" as a feedline?

2I-4.3 What are some advantages in using a parallel conductor feedline?

2I-4.4 What are some disadvantages in using a parallel conductor feedline?

2I-4.5 What kind of antenna feedline is constructed of two conductors maintained a uniform distance apart by insulated spreaders?

# QUESTIONS FOR THE ELEMENT 3
# AMATEUR RADIO OPERATOR EXAMINATION

The questions used in an Element 3 written examination must be taken from the following list. CANDIDATES SHOULD NOT TRY TO MEMORIZE THESE QUESTIONS. The examination is a test of the knowledge a person needs in order to operate an amateur radio station properly. It is not a test of a person's ability to answer questions by rote. See PR Bulletin 1035, Study Guide For FCC Amateur Radio Operator License Examinations for more information.

The questions must be used exactly as stated in this Bulletin. The Volunteer-Examiner Coordinators use their discretion as to the form of the examination answers: essay, single-answer, fill-in-the-blank, multiple-choice, true-false, etc. See Section 97.27 of the Commission's Rules.

Direct any inquiries concerning these questions to the Volunteer-Examiner Coordinator. Do not contact the FCC.

SUBELEMENT 3A - Rules and Regulations (9 questions)

3A-1.1 What is the control point of an amateur station?

3A-1.2 What is the term for the operating position of an amateur station where the control operator function is performed?

3A-2.1 What is an amateur emergency communication?

3A-2.2 What is the term for an amateur radiocommunication directly related to the immediate safety of life of an individual?

3A-2.3 What is the term for an amateur radiocommunication directly related to immediate protection of property?

3A-2.4 Under what circumstances does the FCC declare that a general state of communications emergency exists?

3A-2.5 How does an amateur operator request the FCC to declare that a general state of communications emergency exists?

3A-2.6 What type of instructions are included in an FCC declaration of a general state of communications emergency?

3A-2.7 What should be done by the control operator of an amateur station which has been designated by the FCC to assist in promulgating information relating to a general state of communications emergency?

3A-2.8 During an FCC-declared general state of communications emergency, how must the operation by, and with, amateur stations in the area concerned be conducted?

3A-3.1 Notwithstanding the numerical limitations in the FCC Rules, how much transmitting power shall be used by an amateur radio station?

3A-3.2 What is the maximum transmitting power permitted an amateur station transmitting on 1825-kHz?

3A-3.3 What is the maximum transmitting power permitted an amateur station transmitting on 3725-kHz?

3A-3.4 What is the maximum transmitting power permitted an amateur station transmitting on 7080-kHz?

3A-3.5 What is the maximum transmitting power permitted an amateur station transmitting on 7.125-MHz?

3A-3.6 What is the maximum transmitting power permitted

an amateur station in beacon operation?

3A-3.7 What is the maximum transmitting power permitted an amateur station transmitting on 21.150-MHz?

3A-3.8 What is the maximum transmitting power permitted an amateur station transmitting on 146.52-MHz?

3A-4.1 How must a General control operator at a Novice's station make the station identification?

3A-4.2 How must a newly-upgraded Technician control operator with a Certificate of Successful Completion of Examination identify a station when transmitting on 146.34-MHz pending receipt of a new operator license?

3A-4.3 How must a newly-upgraded General control operator with a Certificate of Successful Completion of Examination identify a station when transmitting on 14.325-MHz pending the receipt of a new operator license?

3A-4.4 When making the station identification by telephony, which language(s) must be used?

3A-4.5 To assist in correct station identification using telephony, what aid does the FCC recommend?

3A-4.6 What emission mode may always be used for station identification, regardless of the transmitting frequency?

3A-5.1 Under what circumstances may a third-party directly participate in radiocommunications from an amateur station?

3A-5.2 Where must the control operator be situated when a third-party is participating in radiocommunications from an amateur station?

3A-5.3 What must the control operator be doing when a third-party is participating in radio-communications from an amateur station?

3A-5.4 Under what circumstances, if any, may a third-party assume the functions of the control operator of an amateur station?

3A-6.1 Under what circumstances, if any, may third-party traffic be transmitted to a foreign country by an amateur station?

3A-6.2 What types of messages may be transmitted by an amateur station to a foreign country for a third-party?

3A-6.3 What types of material compensation, if any, may be involved in third-party traffic transmitted by an amateur station?

3A-6.4 What types of business communications, if any, may be transmitted by an amateur station on behalf of a third-party?

3A-7.1 What kinds of one-way communications by amateur stations are not considered broadcasting?

3A-7.2 What is a one-way communication?

3A-7.3 Under what circumstances may an amateur station transmit a one-way communication consisting of information bulletins?

3A-7.4 What are four types of permissible one-way amateur radiocommunications?

3A-8.1 What are the frequency privileges authorized to a Technician control operator in the HF amateur bands?

3A-8.2 Which operator licenses authorize privileges on the frequency 52.525-MHz?

3A-8.3 Which operator licenses authorize privileges on the frequency 146.52-MHz?

3A-8.4 Which operator licenses authorize privileges on the frequency 223.50-MHz?

3A-8.5 Which operator licenses authorize privileges on the frequency 446.0-MHz?

3A-9.1 What frequency privileges are authorized to the General operator in the 160 meter band?

3A-9.2 What frequency privileges are authorized to the General operator in the 75/80 meter band?

3A-9.3 What frequency privileges are authorized to the General operator in the 40 meter band?

3A-9.4 What frequency privileges are authorized to the General operator in the 30 meter band?

3A-9.5 What frequency privileges are authorized to the General operator in the 20 meter band?

3A-9.6 What frequency privileges are authorized to the General operator in the 15 meter band?

3A-9.7 What frequency privileges are authorized to the General operator in the 12 meter band?

3A-9.8 What frequency privileges are authorized to the General operator in the 10 meter band?

3A-9.9 Which operator licenses authorize privileges on the frequency 1820-kHz?

3A-9.10 Which operator licenses authorize privileges on the frequency 3950-kHz?

3A-9.11 Which operator licenses authorize privileges on the frequency 7230-kHz?

3A-9.12 Which operator licenses authorize privileges on the frequency 10.125-MHz?

3A-9.13 Which operator licenses authorize privileges on the frequency 14.325-MHz?

3A-9.14 Which operator licenses authorize privileges on the frequency 21.425-MHz?

3A-9.15 Which operator licenses authorize privileges on the frequency 24.895-MHz?

3A-9.16 Which operator licenses authorize privileges on the frequency 29.616-MHz?

3A-10.1 On what frequencies within the 80 meter band may F1 emissions be transmitted?

3A-10.2 On what frequencies within the 40 meter band may F1 emissions be transmitted?

3A-10.3 On what frequencies within the 20 meter band may F1 emmissions be transmitted?

3A-10.4 On what frequencies within the 75 meter band may A5 emissions be transmitted?

3A-10.5 On what frequencies within the 20 meter band may A5 emissions be transmitted?

3A-10.6 On what frequencies within the 15 meter band may A4 emissions be transmitted?

3A-10.7 On what frequencies may A1 emissions be transmitted?

3A-11.1 What is the nearest to the band edge an amateur station's transmitting frequency can be set?

3A-11.2 When selecting a transmitting frequency, what allowance should the control operator make for sideband emissions resulting from keying or modulation?

3A-12.1 What is the maximum mean output power an amateur station may use to operate under the special rules for radio control of remote model craft and vehicles?

3A-12.2 What information must be indicated on the writing affixed to the transmitter in order to operate under the special rules for radio control of remote model craft and vehicles?

3A-12.3 What are the station identification requirements for an amateur station operated under the special rules for radio control of remote model craft and vehicles?

3A-12.4 Where must the writing indicating the station call sign and the licensee's name and address be affixed in order to operate under the special rules for radio control of remote craft and vehicles?

3A-13.1 How is the sending speed (signaling rate) for digital communications determined?

3A-13.2 What is the greatest sending speed permitted for an F1 emission on frequencies below 28-MHz?

3A-13.3 What is the greatest sending speed permitted for an F1 emission on frequencies between 28- and 50-MHz?

3A-13.4 What is the greatest sending speed permitted for an F1 emission on frequencies between 50- and 220-MHz?

3A-13.5 What is the greatest sending speed permitted for an F1 emission on frequencies above 220-MHz?

3A-13.6 When A2, F1 or F2 emissions are transmitted on frequencies below 50-MHz, what is the maximum radio or audio frequency shift permitted?

3A-13.7 When A2, F1 or F2 emissions are transmitted on frequencies above 50-MHz, what is the maximum radio or audio frequency shift permitted?

3A-13.8 What is the maximum bandwidth permitted for a transmission from an amateur station using a non-standard digital code on frequencies between 50- and 220-MHz?

3A-13.9 What is the maximum bandwidth permitted for a transmission from an amateur station using a non-standard digital code on frequencies between 220- and 1215-MHz?

3A-13.10 What is the maximum bandwidth permitted for a transmission from an amateur station using a non-standard digital code on frequencies above 1215-MHz?

3A-14.1 What is meant by the term broadcasting?

3A-14.2 What is the only class of station that may be retransmitted by an amateur station?

3A-14.3 Under what circumstances, if any, may a broadcast station retransmit the signals from an amateur station?

3A-14.4 Under what circumstances, if any, may an amateur station be used to engage in some form of broadcasting?

3A-15.1 Under what circumstances, if any, may music be transmitted by an amateur station?

3A-16.1 Under what circumstances, if any, may an amateur station transmit secret codes in order to obscure the meaning of messages?

3A-16.2 What types of abbreviations or signals are not considered codes or ciphers?

3A-16.3 When, if ever, are codes and ciphers permitted in domestic amateur radiocommuniations?

3A-16.4 When, if ever, are codes and ciphers permitted in international amateur radiocommunications?

3A-17.1 Under what circumstances, if any, may amateur stations transmit radiocommunications containing obscene, indecent, or profane words?

SUBELEMENT 3B - Operating Procedures (6 questions)

3B-1.1 What is the meaning of: "Your report is five-seven..."?

3B-1.2 What is the meaning of: "Your report is three-three..."?
3B-1.3 What is the meaning of: "Your report is plus 20dB..."?
3B-1.4 What is meant by the term flattopping in a single-sideband transmission?
3B-1.5 How should the audio gain control be adjusted on a single-sideband transmitter?
3B-1.6 How should the audio gain control be adjusted on a frequency modulated transmitter?
3B-1.7 How should the call sign WE5TZD be stated phonetically?
3B-1.8 How should the call sign KC4HRM be stated phonetically?
3B-1.9 How should the call sign AF6PSQ be stated phonetically?
3B-1.10 How should the call sign NB8LXG be stated phonetically?
3B-2.1 In what segment of the 20 meter band do most amateur RTTY communications take place?
3B-2.2 In what segment of the 80 meter band do most amateur RTTY communications take place?
3B-2.3 What is meant by the term Baudot?
3B-2.4 What is meant by the term ASCII?
3B-2.5 What is meant by the term AMTOR?
3B-2.6 What is the most common frequency shift currently used in RTTY transmissions in the HF amateur bands?
3B-2.7 What is the most common frequency shift currently used on the VHF amateur bands?
3B-2.8 What is an RTTY Mailbox?
3B-2.9 What is the purpose of transmitting a string of RYRYRYRYRYR characters in RTTY?
3B-2.10 What are the two subset modes of AMTOR?
3B-3.1 How should a contact be initiated through a station in repeater operation?
3B-3.2 Why should users of a station in repeater operation pause briefly between transmissions?
3B-3.3 Why should users of a station in repeater operation keep their transmissions short and thoughtful?
3B-3.4 Why should simplex be used where possible instead of using a station in repeater operation?
3B-3.5 What is the proper procedure to break into an on-going QSO through a station in repeater operation?
3B-3.6 What is the purpose of repeater operation in the Amateur Radio Service?
3B-3.7 What is a repeater frequency coordinator?
3B-3.8 What is meant by a bandplan?
3B-3.9 What is the usual input/output frequency separation for a station in repeater operation in the 2 meter band?
3B-3.10 What is the usual input/output frequency separation for a station in repeater operation in the 420-450 MHz band?
3B-3.11 What is the usual input/output frequency separation for a station in repeater operation in the 6 Meter band?
3B-3.12 What is the usual input/output frequency separation for a station in repeater operation in the 220-225 MHz band?
3B-4.1 What is meant by VOX transmitter control?
3B-4.2 What is the common name for the circuit that causes a transmitter to key-on when a person speaks into the microphone?

3B-5.1 What is meant by full break-in telegraphy?

3B-5.2 What Q signal is used to indicate full break-in telegraphy capability?

3B-6.1 When selecting a telegraphy transmitting frequency, what is the minimum frequency separation from a QSO in progress that should be allowed in order to minimize interference?

3B-6.2 When selecting a single-sideband transmitting frequency, what is the minimum frequency separation from a QSO in progress that should be allowed in order to minimize interference?

3B-6.3 When selecting an F1 RTTY transmitting frequency, what is the minimum frequency separation from a QSO in progress that should be allowed in order to minimize interference?

3B-6.4 Why should local amateur communications be conducted on VHF and UHF frequencies?

3B-6.5 How can on-the-air transmissions be minimized during a lengthy transmitter testing or loading up procedure?

3B-6.6 When a frequency conflict arises between a simplex operation and a station in repeater operation, why does good amateur practice call for the simplex operation to move to another frequency?

3B-6.7 What should an amateur operator do before installing a station within one mile of an FCC monitoring station?

3B-6.8 What is the proper Q signal to use to determine whether a frequency is in use before making a transmission?

3B-6.9 What is meant by "making the repeater time out"?

3B-6.10 During the commuter rush hours, which types of operation should relinquish the use of the repeater?

3B-7.1 What is an azimuthal (or great circle) map?

3B-7.2 How is an azimuthal (or great circle) map of importance to an amateur operator conducting international radiocommunications?

3B-7.3 What is the most useful type of map to use when orienting a directional antenna toward a station 5,000 miles distant?

3B-7.4 A directional antenna pointed in the long-path direction to another station is generally oriented how many degrees from the short-path heading?

3B-7.5 What is the short-path beam heading to Antarctica?

3B-8.1 When permitted, transmissions to amateur stations in another country must be limited to only what type of messages?

3B-8.2 In which International Telecommunication Union Region is the continental United States?

3B-8.3 In which International Telecommunication Union Region are Alaska and Hawaii?

3B-8.4 In which International Telecommunication Union Region are American Samoa, Commonwealth of Northern Mariannas Islands, Guam Island, and Wake Island?

3B-8.5 For uniformity in international radiocommunication, what time measurement standard should amateur operators use worldwide?

3B-9.1 What is the proper distress calling procedure when using telephony?

3B-9.2 What is the proper distress calling procedure when using telegraphy?

3B-10.1 What is the Amateur Auxiliary to the FCC's Field Operations Bureau?

3B-10.2 What are the objectives of the Amateur Auxiliary?

SUBELEMENT 3C - Radio Wave Propagation (6 Questions)

3C-1.1 What is the ionosphere?

3C-1.2 Which ionospheric layer limits daytime radiocommunications in the 80 meter band to short distances?

3C-1.3 Why are the electrified characteristics of the ionospheric layers subject to wide variations?

3C-1.4 Which layer of the ionosphere is mainly responsible for long-distance skywave radiocommunications?

3C-1.5 What are the two distinct sub-layers of the F layer during the daytime?

3C-1.6 What is the maximum distance that can be covered in one hop using the F2 layer?

3C-1.7 What is the maximum distance that can be covered in one hop using the E layer?

3C-1.8 What is the lowest region of the ionosphere that is useful for long-distance radio wave propagation?

3C-1.9 What is the average height of maximum ionization of the E layer?

3C-1.10 What is the approximate height of the F2 layer at noontime in the summer?

3C-1.11 What type of solar radiation is most responsible for ionization in the outer atmosphere?

3C-1.12 What is the lowest ionospheric layer?

3C-1.13 What is the critical angle as used in radio wave propagation?

3C-1.14 What is the region of the outer atmosphere which makes long-distance radiocommunications possible as a result of bending of the radio waves?

3C-2.1 Which layer of the ionosphere is most responsible for absorption of radio signals during daylight hours?

3C-2.2 When is ionospheric absorption most pronounced?

3C-2.3 What is the main reason that the 160, 80 and 40 meter bands are useful for short-distance communications during daylight hours?

3C-2.4 What is the principal reason that the 1.8-MHz through 7-MHz bands are useful for short-distance communications only during high-sun hours?

3C-2.5 During daylight hours, what effect does the D layer of the ionosphere have on 80 meter radio waves?

3C-2.6 What causes ionospheric absorption of radio waves?

3C-3.1 What is the highest radio frequency that will be refracted back to earth called?

3C-3.2 What causes the maximum usable frequency to vary?

3C-3.3 If the maximum usable frequency on the path from Minnesota to Africa is 23-MHz, which band should offer the best chance for a successful contact?

3C-3.4 If the maximum usable frequency on the path from Ohio to West Germany is 17-MHz, which band is most favorable for successful communication?

3C-3.5 What does the term maximum usable frequency refer to?

3C-4.1 What is usually the condition of the ionosphere just before sunrise?

3C-4.2 At what time of day does maximum ionization of the ionosphere occur?

3C-4.3 Which two daytime ionospheric layers combine into one layer at night?

3C-4.4 Minimum ionization of the ionosphere occurs daily at what time?

3C-5.1 Over what periods of time do sudden ionospheric disturbances normally last?

3C-5.2 What can be done at an amateur station to continue radiocommmunications during a sudden iononspheric disturbance?

3C-5.3 What effect does a sudden ionospheric disturbance have on the daylight ionospheric propagation of HF radio waves?

3C-5.4 How long does it take a solar disturbance that increases the sun's radiation of charged particles to affect radio wave propagation on Earth?

3C-5.5 How long does it take a solar disturbance that increases the sun's ultraviolet radiation to cause ionospheric disturbances on Earth?

3C-5.6 Sudden ionospheric disturbances occur as a result of radio wave absorption in which layer of the ionosphere?

3C-6.1 When two stations are within each other's skip zone on the frequency being used, what mode of propagation would it be desirable to use?

3C-6.2 What is a characteristic of backscatter signals?

3C-6.3 When is E layer ionization at a maximum?

3C-6.4 What makes backscatter signals often sound distorted?

3C-6.5 What is the radio wave propagation phenomenon that allows a signal to be detected at a distance too far for ground wave propagation but too near for normal sky wave propagation?

3C-6.6 When does ionospheric scatter propagation on the high-frequency bands most often occur?

3C-7.1 What is solar flux?

3C-7.2 What is the solar-flux index?

3C-7.3 When are solar-flux index measurements taken?

3C-7.4 What type of propagation conditions would a solar-flux index value of 66 represent?

3C-7.5 A solar-flux index of 80 would indicate what type of propagation conditions on the 15 meter band?

3C-7.6 A solar-flux index of greater than 80 would indicate what type of propagation conditions on the 10 meter band?

3C-7.7 For widespread long distance openings on the 6 meter band, what solar-flux index values would be required?

3C-7.8 If the MUF is high and HF radiocommunications are generally good for several days, a similar condition can usually be expected how many days later?

3C-8.1 What is the transmission path of a wave that travels directly from the transmitting antenna to the receiving antenna called?

3C-8.2 How are VHF signals within the range of the visible horizon propagated?

3C-9.1 Ducting occurs in which region of the atmosphere?

3C-9.2 What effect does tropospheric bending have on 2 meter radio waves?

3C-9.3 What atmospheric phenomenon causes tropospheric ducting of radio waves?

3C-9.4 Tropospheric ducting occurs as a result of what phenomenon?

3C-9.5 What atmospheric phenomenon causes VHF radio waves to be propagated several hundred miles through stable air masses over oceans?

3C-9.6 In what frequency range does tropospheric ducting most often occur?
3C-10.1 What is a geomagnetic disturbance?
3C-10.2 Which latitude paths are more susceptible to geomagnetic disturbances?
3C-10.3 What can be the effect of a major geomagnetic storm on radiocommunication?

SUBELEMENT 3D - Amateur Radio Practice (9 questions)
3D-1.1 Where should the green wire in an ac line cord be attached in a power supply?
3C-1.2 Where should the colored wire in a three wire line cord (117-vac) be attached in a power supply?
3D-1.3 Where should the white wire in a three wire line cord (117-vac) be attached in a power supply?
3D-1.4 Why is the retaining screw in one terminal of a light socket made of brass while the other one is silver colored?
3D-1.5 Which wires in a four conductor line cord should be attached to fuses in a 234-vac primary (single phase) power supply?
3D-1.6 What size wire is normally used on a 15-ampere, 117-vac household lighting circuit?
3D-1.7 What size wire is normally used on a 20-ampere, 117-vac household appliance circuit?
3D-1.8 What could be a cause of the room lights dimming when the transmitter is keyed?
3D-1.9 What size fuse should be used on a #12 wire household appliance circuit?
3D-2.1 How much electrical current flowing through the human body is usually fatal?
3D-2.2 What is the minimum voltage considered to be dangerous to humans?
3D-2.3 Where should the main power-line switch for a high voltage power supply be situated?
3D-2.4 What safety feature is provided by a bleeder resistor in a power supply?
3D-3.1 What kind of input signal is used to test the amplitude linearity of an SSB transmitter while viewing the output with an oscilloscope?
3D-3.2 To test the amplitude linearity of an SSB transmitter with an oscilloscope, what should the audio input to the transmitter be?
3D-3.3 Why would a two-tone test be used for testing the amplitude linearity of an SSB transmitter?
3D-3.4 What kind of a test is used to check the amplifier portion of an SSB transmitter for amplitude linearity?
3D-3.5 What can be determined by making a "two-tone test" using an oscilloscope?
3D-4.1 How can the grid-current meter in a power amplifier be used as a neutralizing indicator?
3D-4.2 Why is neutralization in some vacuum tube amplifiers necessary?
3D-4.3 How is neutralization of an rf amplifier accomplished?
3D-4.4 For what purpose is a neutralized circuit used in an rf amplifier?
3D-4.5 What is the reason for neutralizing the final amplifier stage of an amateur radio transmitter?
3D-5.1 How can the PEP output of a transmitter be determined with an oscilloscope?
3D-5.2 Where in the antenna transmission line should a peak-reading wattmeter be attached to determine the transmitter power output?

3D-5.3 If a directional rf wattmeter indicates 90 watts forward power and 10 watts reflected power, what is the actual forward power?

3D-5.4 If a directional rf wattmeter indicates 96 watts forward power and 4 watts reflected power, what is the actual forward power?

3D-5.5 What is the approximate dc input power to a Class B power amplifier stage in a transmitter when the emission is F3 and the PEP output is 1500 watts?

3D-5.6 What is the approximate dc input power to a Class C power amplifier in a transmitter when the emission is F1 and the PEP output is 1000 watts?

3D-5.7 What is the approximate dc input power to a Class AB power amplifier stage in a transmitter when the emission is AØ and the PEP output is 500 watts?

3D-6.1 What piece of test equipment contains horizontal and vertical channel amplifiers?

3D-6.2 What types of signals does an oscilloscope measure?

3D-6.3 What is an oscilloscope?

3D-6.4 What can cause phosphor damage to an oscilloscope CRT?

3D-7.1 What is a multimeter?

3D-7.2 How can the range of a voltmeter be extended?

3D-7.3 How is a voltmeter typically connected to a circuit?

3D-7.4 How can the range of an ammeter be extended?

3D-8.1 What is a marker generator?

3D-8.2 What two pieces of test equipment are used to adjust the frequency response of a circuit?

3D-8.3 What type of circuit is used to inject a frequency calibration signal into a communications receiver?

3D-8.4 What does a marker generator do?

3D-8.5 When adjusting a transmitter filter circuit, what device is connected to the transmitter output?

3D-9.1 What is a signal tracer?

3D-9.2 How is a signal tracer used?

3D-9.3 What is a signal tracer normally used for?

3D-10.1 What is the most effective way to reduce or eliminate radio frequency interference to high fidelity systems?

3D-10.2 What should you do if your properly-operating amateur radio equipment is the source of interference on a neighbor's telephone equipment?

3D-10.3 What type of sound would be heard from a public address sytem when audio rectification occurs in response to a nearby single-sideband transmitter?

3D-10.4 How can the possibility of audio rectification occurring be minimized?

3D-10.5 What type of sound would be heard from a public address system when audio rectification occurs in response to a nearby full-carrier AM telephony transmitter?

3D-11.1 What is a reflectometer used for?

3D-11.2 When adjusting the impedance match between an antenna and feedline, where should the match-indicating device be inserted for best accuracy?

3D-11.3 What is the device that indicates impedance mismatches in an antenna system?

3D-11.4 What is a reflectometer?

3D-11.5 Where should a reflectometer be inserted into a long antenna transmission line in order to obtain the most valid standing wave ratio indication?

3D-12.1 What result might be expected when using a single-sideband transmitter even with a properly-adjusted speech processor?

3D-12.2 What is the reason for using a properly-adjusted speech processor with a single-sideband transmitter?

3D-12.3 If a transmitter is 100% modulated, will the use of a speech processor increase the peak power output?

3D-12.4 Under which band conditions should a speech processor not be used?

3D-12.5 What effect can result from using a speech processor with a single-sideband transmitter?

3D-13.1 At what point in the coaxial line should an electronic T-R switch be installed?

3D-13.2 Give a reason for using an electronic T-R switch instead of a mechanical one.

3D-13.3 What station accessory facilitates QSK operation?

3D-14.1 What is a transmatch?

3D-14.2 What is a balanced line?

3D-14.3 What is an unbalanced line?

3D-14.4 What is balun?

3D-14.5 What is the purpose of an antenna matching circuit?

3D-14.6 What is an antenna noise bridge?

3D-14.7 How is an antenna noise bridge used?

3D-14.8 How is a transmatch used?

3D-15.1 How does the emitted waveform from a properly adjusted single-sideband transmitter appear on a monitoring oscilloscope?

3D-15.2 What is the best instrument for checking transmitted signal quality from a telegraphy/single-sideband transmitter?

3D-15.3 What is a monitoring oscilloscope?

3D-15.4 How is a monitoring oscilloscope connected in a station in order to check the quality of the transmitted signal?

3D-16.1 What is a dummy antenna?

3D-16.2 Of what materials may a dummy load suitable for rf be made?

3D-16.3 What station accessory is used in place of an antenna during transmitter tests when no signal radiation is desired?

3D-16.4 What is the purpose of a dummy load?

3D-16.5 A dummy load for use with a 100 watt single-sideband transmitter with 50 ohm output should be able to dissipate at least how many watts?

3D-17.1 What is an S-meter?

3D-17.2 What is the most appropriate instrument to use when determining antenna horizontal radiation patterns?

3D-17.3 What is a field-strength meter?

3D-17.4 What is a simple instrument that can be useful for monitoring relative rf output during antenna and transmitter adjustments?

3D-17.5 When the power output from a transmitter is increased from 250 watts to 1000 watts, how should the S-meter reading on a nearby receiver change?

3D-17.6 How many watts must the power output from a transmitter change to raise the S-meter reading on a nearby receiver from S-7 to S-9?

3D-18.1 For most accurate readings, where should a wattmeter be inserted?

3D-18.2 What factors determine the practical upper frequency limit of an rf wattmeter?

3D-18.3 What is a directional wattmeter?

SUBELEMENT 3E - Electrical Principles (4 questions)

3E-1.1 What is the unit measure of impedance?

3E-1.2 What is the opposition to the flow of an alternating electrical current in a circuit containing both resistance and reactance called?

3E-2.1 What is the unit measure of resistance?

3E-3.1 What is meant by the term reactance?

3E-3.2 In an electrical circuit, what is the opposition to the flow of alternating current caused by inductance or capacitance called?

3E-4.1 On what electrical principle is the transformer based?

3E-4.2 To induce a current in a transformer secondary what must be the condition in the primary?

3E-4.3 Describe the electrical property of an inductor.

3E-5.1 Describe the component parts of a capacitor.

3E-5.2 What is the total capacitance of two equal capacitors, connected in a series?

3E-5.3 What is the total capacitance of two or more capacitors connected in parallel?

3E-6.1 When will a power source deliver maximum output?

3E-7.1 What is the unit measurement of impedance, reactance and resistance called?

3E-7.2 What is an ohm?

3E-7.3 If 120-volts is measured across a 470 ohm resistor, approximately how much current is flowing through the resistor?

3E-8.1 What do the units microfarad and picofarad specify?

3E-8.2 A microfarad equals how many farads?

3E-8.3 A picofarad equals how many farads?

3E-9.1 What is the basic unit of inductance?

3E-10.1 Your signal strength report is "10db over S9". If you turn off your 1-kw amplifier and reduce power to 100 watts, what would the signal strength be?

3E-11.1 How can the current be calculated when the voltage and resistance are known in a dc circuit.

3E-11.2 A 12-volt battery supplies a current of 0.25-amperes to a load. What is the input resistance of this load?

3E-11.3 The product of the current and what force gives the electrical power in a circuit?

3E-12.1 If a 1.0-ampere current source is connected to two parallel-connected 10 ohm resistors, how much current passes through each resistor?

3E-12.2 In a series circuit composed of a voltage source and several resistors, what determines the voltage drop across any particular resistor?

3E-12.3 In a parallel circuit with a voltage source and several branch resistors, what relationship does the total current have to the current in the branch currents?

3E-13.1 If a power supply delivers 200 watts of electrical power at 400-vdc to a load, how much current does the load draw?

3E-13.2 A transmitter's pilot light is connected to 12-vdc and draws 0.2-amperes. How many watts of electrical power are being consumed?

3E-13.3 If a current of 7.0-milliamperes passes through a load resistance of 1.25-kilohms, how many watts are being dissipated?

3E-13.4 How can the power be calculated when the current and voltage are known.

3E-14.1 How is the total resistance found when several

resistors are in series?

3E-14.2 How can the total resistance be calculated in a circuit consisting of two parallel resistors, assuming that each resistor is greater than 3 ohms.

3E-14.3 What is the total inductance of two equal, parallel-connected inductors?

3E-15.1 The primary winding of a transformer has 2250 turns and the secondary winding has 500 turns. If the primary is connected to 117-vac, what is the secondary voltage?

3E-15.2 What is the turns ratio of a transformer to match an audio amplifier having an output impedance of 200 ohms to a speaker having a load impedance of 10 ohms?

3E-15.3 What is the turns ratio of a transformer to match an audio amplifier having an output impedance of 600 ohms to a speaker having a load impedance of 4 ohms?

3E-15.4 A transformer is used to match an audio amplifier having an output impedance of 2000 ohms to a speaker. This transformer has a turns ratio of 24 to 1. What is the impedance of the speaker?

3E-16.1 What is the voltage that would produce the same amount of heat over time in a resistive element as would an applied sine wave ac voltage?

3E-16.2 What is the peak-to-peak voltage of a sine wave which has an RMS voltage of 117-volts?

3E-16.3 A sine wave of 17-volts peak is equivalent to how many volts RMS?

## SUBELEMENT 3F - Circuit Components (3 questions)

3F-1.1 How can you find a carbon resistor's electrical tolerance rating?

3F-1.2 Why would a large size resistor be substituted for a smaller one of the same resistance?

3F-1.3 For what primary purpose is a resistor used in electrical circuits?

3F-1.4 What causes resistors in an electrical circuit to increase in temperature?

3F-1.5 What is the effect of an increase in ambient temperature on a carbon resistor's resistance?

3F-1.6 If the first three color bands on a group of resistors indicate that they all have the same resistance, what further information do you need about each resistor to insure that you can select those of the group that have nearly equal values?

3F-2.1 As the plate area of a capacitor increases, does its maximum possible capacitance decrease, increase or stay the same?

3F-2.2 As the plate spacing of a capacitor increases, what happens to its capacitance?

3F-2.3 What does an electrolytic capacitor contain?

3F-2.4 Where does a capacitor store energy?

3F-2.5 What moves in a typical variable capacitor?

3F-2.6 Are paper dielectric capacitors polarized?

3F-2.7 What kind of capacitors do most power-supply filters use?

3F-2.8 What type of capacitor is most often used in power supply circuits to "filter" the rectified alternating current?

3F-2.9 What type of capacitor is used in power supply circuits to filter out transient voltage spikes across the transformer secondary winding.

3F-2.10 How are the characteristics of a capacitor usually specified?

3F-3.1 A small air-core coil has an inductance of 5-microhenrys. What do you have to do if you want a 5-millihenry coil with the same physical dimensions?

3F-3.2 Describe an inductor.

3F-3.3 As an iron core is inserted in a coil, what happens to its inductance?

3F-3.4 For radio frequency power applications, with which type of inductor would you get the least amount of loss?

3F-3.5 How may inductors become self-resonant?

3F-3.6 Where does an inductor store energy?

3F-3.7 What electrical circuit component can change a 120-vac to a 400-vac?

3F-4.1 In a transformer, what is the source of energy connected to?

3F-4.2 When no load is attached to the secondary winding of a transformer, what is current in the primary winding called?

3F-4.3 In what terms are the primary and secondary winding ratings of a power transformer usually specified?

3F-5.1 What is the peak-inverse-voltage rating of a power-supply rectifier?

3F-5.2 Why must silicon rectifier diodes be thermally protected?

3F-5.3 What is a heat sink?

3F-5.4 Silicon diode rectifiers of the type used in a power-supply circuit have two major ratings which must not be exceeded. What are these two ratings?

SUBELEMENT 3G - Practical Circuits (2 questions)

3G-1.1 Why should a resistor and capacitor be wired in parallel with power-supply rectifier diodes?

3G-1.2 What function do capacitors serve when resistors and capacitors are connected in parallel with high voltage power-supply rectifier diodes?

3G-1.3 Describe the output waveform of an unfiltered full wave rectifier connected to a resistive load.

3G-1.4 How many degrees of a cycle does a half-wave rectifier utilize?

3G-1.5 How many degrees of a cycle does a full-wave rectifier utilize?

3G-1.6 Where is a power supply "bleeder resistor" connected?

3G-1.7 What components comprise a power supply filter network?

3G-1.8 What should a power supply rectifier's peak-inverse-voltage rating be in either a full-wave or half-wave rectifier circuit?

3G-2.1 What is a high-pass filter usually connected to?

3G-2.2 Where is the proper place to install a high-pass filter?

3G-2.3 Where is a band-pass filter usually installed?

3G-2.4 Does a low-pass filter attenuate electrical energy below or above its cut-off frequency?

3G-2.5 What is a circuit called which passes electrical energy above a certain frequency, but blocks electrical energy below that frequency?

3G-2.6 What is a circuit called which passes electrical energy below a certain frequency, but blocks electrical energy above that frequency?

3G-2.7 What is a circuit called that blocks electrical energy above a certain frequency and also blocks electrical energy below a lower frequency, but passes electrical energy between the two frequencies?

3G-2.8 For proper operation, what should the impedance of a low-pass filter be as compared to the impedance of the transmission line into which it is inserted.

3G-2.9 What general group of radio frequency energy does a band-pass filter reject?

3G-3.1 Name a circuit that is likely to be found in all types of receivers.

3G-3.2 In a filter-type SSB transmitter, what stage combines radio frequency and audio frequency energy to produce a double-sideband suppressed carrier signal?

3G-3.3 In a superheterodyne receiver intended for AM reception, what stage combines the received radio frequencies with energy from a local oscillator to produce an output operating at the receiver's intermediate frequency?

SUBELEMENT 3H - Signals and Emissions (4 questions)

3H-1.1 What emission type is AØ?

3H-1.2 What emission type is A3?

3H-1.3 What emission type is A3J?

3H-1.4 What emission type is F1?

3H-1.5 What emission type is F2?

3H-1.6 What emission type is F3?

3H-1.7 What is the emission symbol for telegraphy by frequency shift keying without the use of a modulating audio frequency?

3H-1.8 What is the emission symbol for telegraphy by the on-off keying of a frequency modulating audio frequency?

3H-1.9 What is the emission symbol for telephony by amplitude modulation?

3H-1.10 What is the emission symbol for telephony by frequency modulation?

3H-2.1 What is altering the amplitude, phase or frequency of a radio frequency wave for the purpose of conveying information called?

3H-3.1 In what type of modulation does the instantaneous amplitude (envelope) of the radio frequency signal vary in accordance with the modulating audio frequency signal?

3H-3.2 What determines the spectrum space occupied by each group of sideband frequencies generated by a correctly operating amplitude modulated transmitter?

3H-4.1 What does suppressing the carrier in an AM signal change the emission type to?

3H-4.2 What is one advantage of double-sideband suppressed-carrier transmission over standard full-carrier AM?

3H-5.1 Which one of the popular voice-modulated emissions normally has the narrowest bandwidth?

3H-5.2 What type of emission is produced by a radiotelephone transmitter that uses a balanced modulator followed by a 2.5-kHz bandpass filter?

3H-6.1 What feature makes the F3 emission mode especially suitable for channelized local VHF/UHF communication?

3H-6.2 What type of emission is produced by a transmitter

that uses a reactance modulator stage?

3H-7.1 What other modulation system does phase modulation most resemble?

3H-7.2 If a reactance modulator is connected to an rf power amplifier, what will the predominantly resulting emission be?

3H-8.1 What purpose does the carrier serve in an A3 emission?

3H-8.2 What signal component appears in the center of an amplitude modulated transmitter's emitted bandwidth?

3H-9.1 What sideband frequencies will be generated by an AM transmitter having a carrier frequency of 7250-kHz when it is modulated less than 100 percent by an 800-Hz pure sine wave?

3H-9.2 Which emission type does not have sidebands resulting from modulation?

3H-10.1 How many times over the maximum deviation is the bandwidth of an F3 transmitter's emissions?

3H-10.2 Is the bandwidth of a VHF FM transmitter's emissions narrower or greater at the operating frequency than it is at the output of the modulated oscillator operating at HF?

3H-11.1 When viewing an amplitude modulated transmitter's rf envelope on an oscilloscope, what affects the envelope's shape during modulation?

3H-12.1 To what is the frequency deviation of an FM emission proportional?

3H-13.1 What usually results if a voice transmitter is overmodulated?

3H-13.2 What happens if an A3 emission is overmodulated?

3H-14.1 What is the result of overdeviation of the oscillator in an F3 transmitter?

3H-14.2 What causes splatter?

3H-15.1 An F3 transmitter is to have 5-kHz deviation at a final output frequency of 146.52-MHz. The reactance-modulated oscillator frequency is 12.21-MHz and the final frequency is obtained through multiplication. What is the oscillator frequency deviation?

3H-15.2 The input to a particular stage in an amateur radio transmitter is a signal centered on the frequency 5.3 MHz. The output from this stage contains the same signal translated in frequency in such a way that it is centered on 14.3 MHz. What type of stage is it?

3H-16.1 What type of emission can be used for radioteleprinting?

3H-16.2 What are the two states of the teleprinter codes most commonly used by amateur radio operators?

3H-16.3 What emission type results when an audio frequency shift keyer is connected to the microphone input of a properly adjusted SSB transmitter?

3H-16.4 How many frequency components are in the signal from an audio frequency shift keyer at any instant?

3H-16.5 How is frequency shift related to keying speed in an fsk emission?

SUBELEMENT 3I - Antennas and Feedlines (7 questions)

3I-1.1 What kind of antenna best reduces the received signal strength of signals coming from some directions while strengthening signals arriving from a certain desired direction?

3I-1.2 What is a Yagi antenna?

3I-1.3 Why is a Yagi antenna often used for amateur radiocommunications on the 20 meter band?

3I-1.4 Give a physical description of the radiating elements of a horizontally-polarized Yagi antenna.

3I-1.5 What is the name of a parasitic beam antenna using two or more straight metal-tubing elements arranged physically parallel to each other?

3I-1.6 How many directly-driven elements does a Yagi antenna have?

3I-1.7 What sort of matching system is best suited to match unbalanced coaxial feedlines to Yagi antennas when no balun is being used?

3I-1.8 What is meant by the term parasitic when describing beam antennas?

3I-1.9 What is the correct method of increasing the bandwidth of a parasitic beam antenna?

3I-2.1 How much gain can a two-element quad antenna be expected to provide over a one-half wavelength dipole?

3I-2.2 What kind of antenna array is composed of a square or diamond-shaped full-wave closed loop driven element with parallel parasitic element(s)?

3I-2.3 Approximately how long is one side of the driven element of a quad antenna?

3I-2.4 Approximately how long is the wire in the driven element of a quad antenna?

3I-2.5 What is a delta loop?

3I-2.6 Give a brief description of the radiating elements in a quad antenna?

3I-2.7 In quad antennas, how does the electrical length of the director element compare with that of the driven element?

3I-2.8 In quad antennas, how does the electrical length of the reflector element compare with that of the driven element?

3I-2.9 How would you modify a 20 meter two-element quad antenna to provide 15 and 10 meter coverage?

3I-3.1 A square one-wavelength quad loop is to resonate at 21,250-kHz. How long should each side be?

3I-3.2 How tall must a tower be to support a 20 meter Yagi antenna one-half wavelength above a perfectly conducting surface?

3I-3.3 What length should the radials be on a ground-plane antenna intended for use in the 30 meter band?

3I-3.4 Approximately how long is each element of a delta-loop antenna?

3I-3.5 A square one-wavelength loop is to be resonant at 14.175-MHz. How long should each side be?

3I-3.6 State the formula for approximating the full-wave driven element loop lengths for HF quad antennas.

3I-3.7 State the formula for approximating the full-wave reflector loop length for HF quad antennas.

3I-3.8 State the formula for approximating the full-wave director loop lengths for HF quad antennas.

3I-3.9 How long should each side of a square one-wavelength loop be for resonance on 21.225-MHz?

3I-4.1 What kind of radio waves does a half-wavelength antenna that is perpendicular to the earth's surface radiate?

3I-4.2 Does most man-made electrical noise radiation in the HF-VHF spectrum tend to be vertically or horizontally polarized?

3I-4.3 What do the terms vertical and horizontal as applied to wave polarization refer to?

3I-4.4 Are vertical and horizontal wave polarization examples of elliptical or linear polarization?

3I-4.5 What electromagnetic polarization does a cubical quad antenna have that has top and bottom sides running parallel to the earth, and the feedpoint midway in the driven element's bottom side?

3I-4.6 What electromagnetic polarization does a cubical quad antenna have that has top and bottom sides running parallel to the earth, and the feedpoint midway up the driven element's vertical side?

3I-4.7 What electromagnetic polarization does a cubical quad antenna have that is set in a diamond configuration with all sides at 45 degrees to the earth's surface, and with the driven element fed at the bottom corner?

3I-4.8 What electromagnetic polarization does a cubical quad antenna have that is set in a diamond configuration with all sides at 45 degrees to the earth's surface, and with the driven element fed at the side corner?

3I-5.1 What is the feedpoint impedance for a half-wave dipole antenna suspended horizontally one-quarter wavelength or more above the ground?

3I-5.2 What is the feedpoint impedance for a quarter wavelength vertical antenna with a horizontal ground plane?

3I-6.1 When compared to a dipole, what directional radiation characteristics does a quad antenna have?

3I-6.2 What radiation pattern does an ideal half-wave dipole have?

3I-6.3 How does proximity to the ground affect the radiation pattern of a horizontal dipole antenna?

3I-6.4 What is meant by antenna front-to-back ratio?

3I-6.5 If a slightly smaller parasitic element is placed parallel to a dipole antenna, a few feet away from it, with both elements in the same horizontal plane, what effect will this have on the radiation pattern of the antenna?

3I-6.6 Define main lobe as used in directional antenna theory.

3I-6.7 What is a directional antenna?

3I-7.1 Upon what does the characteristic impedance of a parallel conductor antenna feedline depend?

3I-7.2 What is the characteristic impedance of various coaxial cables commonly used at amateur radio stations for antenna transmission lines?

3I-7.3 What effect, if any, does the length have upon the characteristic impedance of a coaxial cable?

3I-7.4 What is the characteristic impedance of common TV-type twin lead?

3I-8.1 What is standing wave ratio?

3I-8.2 When is corona likely to occur with a parallel conductor feedline?

3I-9.1 What is standing wave ratio a measure of?

3I-9.2 When an unbalanced feedline's SWR increases, what happens to the power loss in that feedline?

3I-9.3 A resonant antenna having a feedpoint impedance of 200 ohms resistance is connected to a feedline having a characteristic impedance of 50 ohms. What will the standing wave ratio of this system be?

3I-9.4 When a feedline is terminated by a resistance equal to its characteristic impedance, how is the feedline's power loss affected?

3I-10.1 Why is a split-stator capacitor useful in a matching circuit?

3I-10.2 How is an inductively-coupled matching network used in an antenna system using a center-fed resonant dipole and coaxial feedline?

3I-10.3 What kind of 1:1 ratio transformer is installed in transmitting feedlines to couple coaxial cable to a balanced antenna?

3I-11.1 How is the amount of attenuation of a radio signal passing through a coaxial cable affected by the characteristic impedance of the cable?

3I-11.2 How does the amount of attenuation of a 2 meter radio frequency signal passing through a coaxial cable differ than one at 160 meters?

3I-11.3 What type of feedline is best suited to operating at a high standing wave ratio?

3I-11.4 What is the effect on feedline attenuation when flat brown "twin-lead" gets wet?

3I-11.5 What is the general relationship between feedline loss and frequency?

3I-11.6 What happens to radio frequency energy not delivered to the antenna by a lossy coaxial feedline?

3I-11.7 Why might silicone grease or automotive car wax be applied to the outside of flat ribbon twin-lead feedline?

3I-11.8 In what values is radio frequency feedline loss usually expressed?

3I-11.9 As operating frequency decreases, what happens to conductor loss in a feedline?

3I-11.10 As operating frequency increases, what happens to dielectric loss in a feedline?

3I-11.11 As operating frequency increases, what happens to conductor loss in a feedline?

3I-11.12 As operating frequency decreases, what happens to dielectric loss in a feedline?

3I-12.1 What condition must be satisfied to prevent standing waves of voltage and current on an antenna feedline?

# QUESTIONS FOR THE ELEMENT 4A AMATEUR RADIO OPERATOR LICENSE EXAMINATION

The questions used in an Element 4A written examination must be taken from the following list. CANDIDATES SHOULD NOT TRY TO MEMORIZE THESE QUESTIONS! The examination is a test of the knowledge a person needs in order to operate an amateur radio station properly. It is not a test of a person's ability to answer questions by rote. See PR Bulletin 1035, Study Guide For The Amateur Radio Operator Examinations for more information.

The questions must be used exactly as stated in this Bulletin. The Volunteer-Examiner Coordinators may use their discretion as to the form of the examination answers. They may be essay, multiple-choice or single answer type. See Section 97.27 of the Commission's Rules.

Direct any inquiries concerning these questions to the Volunteer-Examiner Coordinator. Do not contact the FCC.

SUBELEMENT 4A - Rules and Regulations (6 questions)

4AA-1.1 What are the frequency privileges authorized to the Advanced class operator in the 75/80 meter band?

4AA-1.2 What are the frequency privileges authorized to the Advanced class operator in the 40 meter band?

4AA-1.3 What are the frequency privileges authorized to the Advanced class operator in the 20 meter band?

4AA-1.4 What are the frequency privileges authorized to the Advanced class operator in the 15 meter band?

4AA-2.1 What is meant by automatic retransmission?

4AA-2.2 What is the term for the retransmission of signals by an amateur radio station whereby the retransmitting station is actuated solely by the presence of a received signal through electrical or electromechanical means, i.e., without any direct, positive action by the control operator?

4AA-2.3 Under what circumstances, if any, may an amateur radio station automatically retransmit programs or the radio signals of other amateur radio stations?

4AA-2.4 What is meant by manual retransmission?

4AA-3.1 What is meant by repeater operation?

4AA-3.2 What is a closed repeater?

4AA-3.3 What frequencies are authorized for stations in repeater operation?

4AA-3.4 What determines the maximum effective radiated power a station in repeater operation may use?

4AA-3.5 How is effective radiated power determined?

4AA-4.1 What is meant by auxiliary operation?

4AA-4.2 Give three uses for stations in auxiliary operation.

4AA-4.3 A station in auxiliary operation may only communicate with which stations?

4AA-4.4 What frequencies are authorized for stations in auxiliary operation?

4AA-5.1 What is meant by remote control of an amateur radio station?

4AA-5.2 How do the responsibilities of the control operator of a station under remote control differ from one under local control?

4AA-5.3 If the control link for a station under remote control malfunctions, how long may the station contirue to transmit?

4AA-5.4 What frequencies are authorized for radio remote control of an amateur radio station?

4AA-5.5 What frequencies are authorized for radio remote control of a station in repeater operation?

4AA-6.1 What is meant by automatic control of an amateur radio station?

4AA-6.2 How do the responsibilities of the control operator of a station under automatic control differ from one under local control?

4AA-6.3 Which amateur radio stations may be operated by automatic control?

4AA-7.1 What is meant by control link?

4AA-7.2 What is the term for apparatus to effect remote control between a control point and a remotely controlled station?

4AA-8.1 What is a system network diagram?

4AA-8.2 What type of diagram shows each station and its relationship to other stations in a network of stations, and to the control point(s)?

4AA-9.1 At what level of modulation must a station in repeater operation or one in auxiliary operation transmit its identification?

4AA-9.2 What additional station identification requirements apply to stations in repeater operation?

4AA-9.3 What additional station identification requirements apply to stations in auxiliary operation?

4AA-10.1 When is prior FCC approval required before constructing or altering an antenna structure?

4AA-10.2 What must an amateur radio operator obtain from the FCC before constructing or altering an antenna structure more than 200 feet high?

4AA-11.1 How is antenna height above average terrain determined?

4AA-11.2 For a station in repeater operation transmitting on 146.94 MHz, what is the maximum ERP permitted for an antenna height above average terrain above 1050 feet?

4AA-12.1 What is meant by business communications?

4AA-12.2 What is the term for a transmission or communication the purpose of which is to facilitate the regular business or commercial affairs of any party?

4AA-12.3 Under what conditions, if any, may business communications be transmitted by an amateur radio station?

4AA-13.1 What are the only types of messages that may be transmitted to amateur radio stations in a foreign country?

4AA-13.2 What are the limitations on international communications regarding the types of messages transmitted?

4AA-14.1 Under what circumstances, if any, may amateur radio operators accept payment for using their stations to send messages?

4AA-14.2 Under what circumstances, if any, could the licensee of a station in repeater operation accept remuneration for providing communication services to another party?

4AA-15.1 Who is responsible for preparing an Element 1(A)

telegraphy examination?

4AA-15.2 What must the Element 1(A) telegraphy examination prove?

4AA-15.3 Which telegraphy characters are used in an Element 1(A) telegraphy examination?

4AA-16.1 Who is responsible for preparing an Element 2 written examination?

4AA-16.2 Where do volunteer examiners obtain the questions for preparing an Element 2 written examination?

4AA-17.1 Who is eligible for administering an examination for the Novice class operator license?

4AA-17.2 For how long must the volunteer examiner for a Novice class examination retain the test papers?

4AA-17.3 Where must the volunteer examiner for a Novice class examination retain the test papers?

4AA-18.1 What is the passing score on a written examination element for Novice class?

4AA-18.2 For a 20 question Element 2 written examination, how many correct answers constitute a passing score?

4AA-18.3 In a telegraphy examination, how many characters are counted as one word?

4AA-19.1 What is the minimum age to be a volunteer examiner?

4AA-19.2 Under what circumstances, if any, may a volunteer examiner be compensated for his/her services?

4AA-19.3 May a person whose amateur radio station license or amateur radio operator license has ever been revoked or suspended be a volunteer examiner?

4AA-19.4 Under what circumstances, if any, may a person who is an employee of a company which is engaged in the distribution of equipment used in connection with amateur radio transmissions be a volunteer examiner?

4AA-20.1 What are the penalties for fraudulently giving examinations?

4AA-20.2 What are the penalties for giving examinations for money or other considerations?

## SUBELEMENT 4AB - Amateur Radio Practice (1 question)

4AB-1.1 What is facsimile?

4AB-1.2 What is the scan rate for a facsimile picture?

4AB-1.3 What is the transmission time for facsimile?

4AB-1.4 What is the term for the transmission of printed pictures?

4AB-1.5 What is A4 type emission?

4AB-1.6 What is slow-scan television?

4AB-1.7 What is the scan rate commonly used for amateur slow-scan television?

4AB-1.8 How many lines are there in each frame of a slow-scan television picture?

4AB-1.9 What is the audio frequency for black in a slow-scan television picture?

4AB-1.10 What is the audio frequency for white in a slow-scan television picture?

## SUBELEMENT 4AC - Radio Wave Propagation (2 questions)

4AC-1.1 What is a sporadic-E condition?

4AC-1.2 What is the propagation condition called where

scattered patches of relatively dense ionization develops seasonally at E layer heights?

4AC-1.3 In what region of the world is sporadic-E most prevalent?

4AC-1.4 On which amateur frequency band is extended distant propagation effect of sporadic-E propagation most often observed?

4AC-1.5 What appears to be the major cause of the sporadic-E condition?

4AC-2.1 What is a selective fading effect?

4AC-2.2 What is the propagation effect called when phase differences between radio wave components of the same transmission are experienced at the recovery station?

4AC-2.3 What is the major cause of selective fading?

4AC-2.4 Which emission modes suffer the most from selective fading?

4AC-2.5 How does the bandwidth of the transmitted signal affect selective fading?

4AC-3.1 What effect does auroral activity have upon radio communications?

4AC-3.2 What is the cause of auroral activity?

4AC-3.3 In the northern hemisphere, in which direction should a directional antenna be pointed to take maximum advantage of auroral propagation?

4AC-3.4 Where in the ionosphere does auroral activity occur?

4AC-3.5 Which emission modes are best for auroral propagation?

4AC-4.1 Why does the radio-path horizon distance exceed the geometric horizon?

4AC-4.2 How much further does the radio-path horizon distance exceed the geometric horizon?

4AC-4.3 To what distance is VHF propagation ordinarily limited?

4AC-4.4 What propagation condition is usually indicated when a VHF signal is received from a station over 500 miles away?

4AC-4.5 What happens to a radio wave as it travels in space and collides with other particles?

SUBELEMENT 4AD - Amateur Radio Practice (4 questions)

4AD-1.1 What is a frequency standard?

4AD-1.2 What is a frequency-marker generator?

4AD-1.3 How is a frequency-marker generator used?

4AD-1.4 What is a frequency counter?

4AD-1.5 How is a frequency counter used?

4AD-1.6 A frequency counter indicates a transmit frequency of 146,520,000 Hz. The counter has a time base accuracy of +/- 1.0 ppm. What is the most the actual transmitter frequency could differ from the counter reading?

4AD-1.7 A frequency counter indicates a transmit frequency of 146,520,00 Hz. The counter has a time base accuracy of +/- 0.1 ppm. What is the most the actual transmitter frequency could differ from the counter reading?

4AD-1.8 A frequency counter indicates a transmit frequency of 146,520,000 Hz. The counter has a time base accuracy of +/- 10 ppm. What is the most the actual transmitter frequency could differ from the counter reading?

4AD-1.9 A frequency counter indicates a transmit

frequency of 432,100,000 Hz. The counter has a time base accuracy of +/- 1.0 ppm. What is the most the actual transmitter frequency could differ from the counter reading?

4AD-1.10 A frequency counter indicates a transmit frequency of 432,100,000 Hz. The counter has a time base accuracy of +/- 0.1 ppm. What is the most the actual transmit frequency could differ from the counter reading?

4AD-1.11 A frequency counter indicates a transmit frequency of 432,100,000 Hz. The counter has a time base accuracy of +/- 10 ppm. What is the most the actual transmit frequency could differ from the counter reading?

4AD-2.1 What is a dip-meter?

4AD-2.2 Why is a dip-meter used by many amateur operators?

4AD-2.3 How does a dip-meter work?

4AD-2.4 Name three ways in which an amateur could use a dip-meter.

4AD-2.5 What types of coupling occur between a dip-meter and a tuned circuit being checked?

4AD-2.6 For best accuracy when using a dip-meter, how tight should it be coupled with the tuned circuit being checked?

4AD-2.7 What happens in a dip-meter when it is too tightly coupled with the tuned circuit being checked?

4AD-3.1 What factors limit the accuracy, frequency response, and stability of an oscilloscope?

4AD-3.2 What factors limit the accuracy, frequency response, and stability of a D'Arsonval movement type meter?

4AD-3.3 What factors limit the accuracy, frequency response, and stability of a frequency counter?

4AD-3.4 How can the frequency response of an oscillosope be improved?

4AD-3.5 How can the accuracy of a frequency counter be improved?

4AD-4.1 What is the condition called which occurs when the signals of two transmitters in close proximity mix together in one or both of their final amplifiers, and unwanted signals at the sum and difference frequencies of the original transmissions are generated?

4AD-4.2 How does intermodulation interference between two transmitters usually occur?

4AD-4.3 How can intermodulation interference between two transmitters in close proximity often be reduced or eliminated?

4AD-4.4 What can occur when a non-linear amplifier is used with a single sideband transmitter?

4AD-4.5 How can even-order harmonics be reduced or prevented in transmitter amplifier design?

4AD-5.1 What is receiver desensitizing?

4AD-5.2 What is the term used to refer to the reduction of receiver gain caused by the signals of a nearby station transmitting in the same frequency band?

4AD-5.3 What is the term used to refer to a reduction in receiver sensitivity caused by unwanted high-level adjacent channel signals?

4AD-5.4 What causes receiver desensitizing?

4AD-5.5 How can receiver desensitizing be reduced?

4AD-6.1 What is cross-modulation interference?

4AD-6.2 What is the term used to refer to the condition where the signals from a very strong station are superimposed on other signals being received?

4AD-6.3 How can cross-modulation in a receiver be reduced?

4AD-6.4 What is result of cross-modulation?

4AD-7.1 What is the capture-effect?

4AD-7.2 What is the term used to refer to the reception blockage of one particular FM signal by another stronger FM signal?

4AD-7.3 With which modulation type is the capture-effect most pronounced?

## SUBELEMENT 4AE - Electrical Principles (10 questions)

4AE-1.1 What is reactive power?

4AE-1.2 What is the term for an out-of-phase, non-productive power associated with inductors and capacitors?

4AE-1.3 What is the term for energy that is stored in an electromagnetic or electrostatic field?

4AE-1.4 What is responsible for the phenomenon when voltages across reactances in series can often be larger than the voltages applied to them?

4AE-2.1 What is resonance in an electrical circuit?

4AE-2.2 Under what conditions does resonance occur in an electrical circuit?

4AE-2.3 What is the term for the phenomena which occurs in an electrical circuit when the inductive reactance equals the capacitive reactance?

4AE-2.4 What is the approximate magnitude of the impedance of a series R-L-C circuit at resonance?

4AE-2.5 What is the approximate magnitude of the impedance of a parallel R-L-C circuit at resonance?

4AE-2.6 What is the characteristic of the current flow in a series R-L-C circuit at resonance?

4AE-2.7 What is the characteristic of the current flow in a parallel R-L-C circuit at resonance?

4AE-3.1 What is the skin effect?

4AE-3.2 What is the term for the phenomenon where most of a radio frequency current flows along the surface of the conductor?

4AE-3.3 At radio frequencies, where does practically all of the current flow in a conductor?

4AE-3.4 Why does practically all of a radio frequency current flow within a few thousandths-of-an-inch of the conductor's surface?

4AE-3.5 Why is the resistance of a conductor different for radio frequency alternating current than for direct current?

4AE-4.1 What is a magnetic field?

4AE-4.2 In what direction is the magnetic field about a conductor in which there is a current flowing?

4AE-4.3 What device is used to store electrical energy in an electrostatic field?

4AE-4.4 What is the term used to express the amount of electrical energy stored in an electrostatic field?

4AE-4.5 What factors determine the capacitance of a capacitor?

4AE-4.6 What is the dialectric constant for air?

4AE-4.7 What determines the strength of the magnetic field around a conductor?

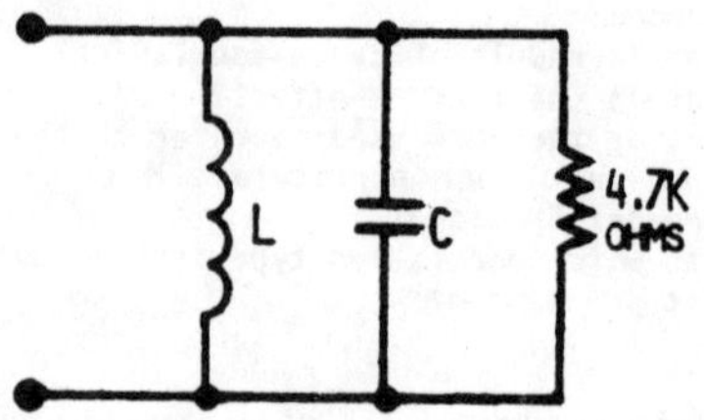

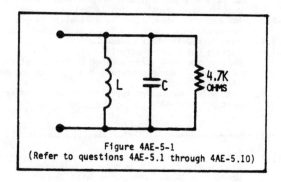

Figure 4AE-5-1
(Refer to questions 4AE-5.1 through 4AE-5.10)

4AE-5.1 What is the resonant frequency of the circuit shown in Figure 4E-5-1 when:
L = 50 microhenries and C = 40 picofarads?
4AE-5.2 When L = 40 microhenries and C = 200 picofarads?
4AE-5.3 When L = 50 microhenries and C = 10 picofarads?
4AE-5.4 When L = 25 microhenries and C = 10 picofarads?
4AE-5.5 When L = 3 microhenries and C = 40 picofarads?
4AE-5.6 When L = 4 microhenries and C = 20 picofarads?
4AE-5.7 When L = 8 microhenries and C = 7 picofarads?
4AE-5.8 When L = 3 microhenries and C = 15 picofarads?
4AE-5.9 When L = 4 microhenries and C = 8 picofarads?
4AE-5.10 When L = 1 microhenries and C = 9 picofarads?

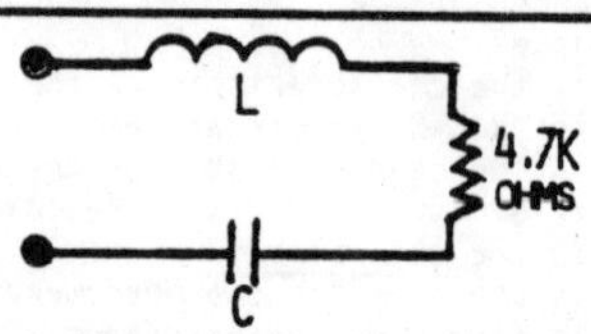

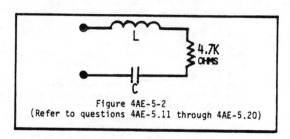

Figure 4AE-5-2
(Refer to questions 4AE-5.11 through 4AE-5.20)

4AE-5.11 What is the resonant frquency of the circuit shown in Figure 4AE-5-2 when:
L = 1 microhenries and C = 10 picofarads?
4AE-5.12 When L = 2 microhenries and C = 15 picofarads?
4AE-5.13 When L = 5 microhenries and C = 9 picofarads?
4AE-5.14 When L = 2 microhenries and C = 30 picofarads?
4AE-5.15 When L = 15 microhenries and C = 5 picofarads?
4AE-5.16 When L = 3 microhenries and C = 40 picofarads?
4AE-5.17 When L = 40 microhenries and C = 6 picofarads?
4AE-5.18 When L = 10 microhenries and C = 50 picofarads?
4AE-5.19 When L = 200 microhenries and C = 10 picofarads?
4AE-5.20 When L = 90 microhenries and C = 100 picofarads?
4AE-5.21 What is the half-power bandwidth of a parallel resonant circuit which has a resonant frequency of 1.8 MHz and a Q of 95?
4AE-5.22 What is the half-power bandwidth of a parallel

resonant circuit which has a resonant frequency of 3.6 MHz and a Q of 218?

4AE-5.23 What is the half-power bandwidth of a parallel resonant circuit which has a resonant frequency of 7.1 MHz and a Q of 150?

4AE-5.24 What is the half-power bandwidth of a parallel resonant circuit which has a resonant frequency of 12.8 MHz and a Q of 218?

4AE-5.25 What is the half-power bandwidth of a parallel resonant circuit which has a resonant frequency of 14.25 MHz and a Q of 150?

4AE-5.26 What is the half-power bandwidth of a parallel resonant circuit which has a resonant frequency of 21.15 MHz and a Q of 95?

4AE-5.27 What is the half-power bandwidth of a parallel resonant circuit which has a resonant frequency of 10.1 MHz and a Q of 225?

4AE-5.28 What is the half-power bandwidth of a parallel resonant circuit which has a resonant frequency of 18.1 MHz and a Q of 195?

4AE-5.29 What is the half-power bandwidth of a parallel resonant circuit which has a resonant frequency of 3.7 MHz and a Q of 118?

4AE-5.30 What is the half-power bandwidth of a parallel resonant circuit which has a resonant frequency of 14.25 MHz and a Q of 187?

4AE-5.31 What is the Q of the circuit shown in Figure 4AE-5-3 when the resonant frequency is 14.128 MHz, the inductance is 2.7 microhenries and the resistance is 18,000 ohms?

4AE-5.32 When the resonant frequency is 14.128 MHz, the inductance is 4.7 microhenries and the resistance is 18,000 ohms?

4AE-5.33 When the resonant frequency is 4.468 MHz, the inductance is 47 microhenries and the resistance is 180 ohms?

4AE-5.34 When the resonant frequency is 14.225 MHz, the inductance is 3.5 microhenries and the resistance is 10,000 ohms?

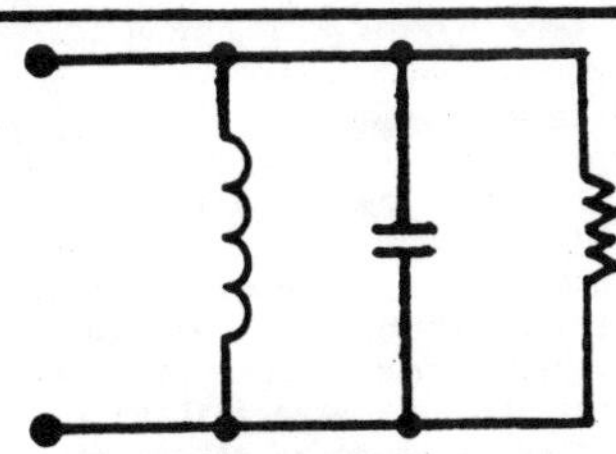

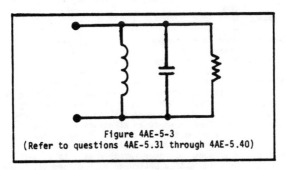

Figure 4AE-5-3
(Refer to questions 4AE-5.31 through 4AE-5.40)

4AE-5.35 When the resonant frequency is 7.125 MHz, the inductance is 8.2 microhenries and the resistance is 1,000 ohms?

4AE-5.36 When the resonant frequency is 7.125 MHz, the inductance is 10.1 microhenries and the resistance is 100 ohms?

4AE-5.37 When the resonant frequency is 7.125 MHz, the inductance is 12.6 microhenries and the resistance is 22,000 ohms?

4AE-5.38 When the resonant frequency is 3.625 MHz, the inductance is 3 microhenries and the resistance is 2,200 ohms?

4AE-5.39 When the resonant frequency is 3.625 MHz, the inductance is 42 microhenries and the resistance is 220 ohms?

4AE-5.40 When the resonant frequency is 3.625 MHz, the inductance is 43 microhenries and the resistance is 1,800 ohms?

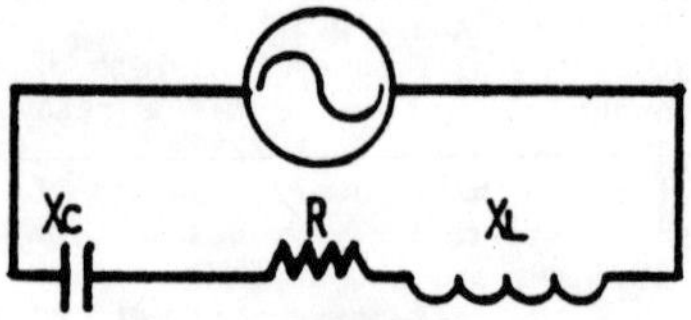

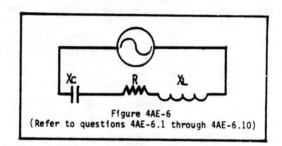

Figure 4AE-6
(Refer to questions 4AE-6.1 through 4AE-6.10)

4AE-6.1 What is the phase angle between the voltage across, and the current through, the circuit in Figure 4AE-6, when Xc = 25 ohms, R = 100 ohms, and Xl = 100 ohms?

4AE-6.2 When Xc = 25 ohms, R = 100 ohms, and Xl = 50 ohms?

4AE-6.3 When Xc = 500 ohms, R = 1000 ohms, and Xl = 250 ohms?

4AE-6.4 When Xc = 75 ohms, R = 100 ohms, and Xl = 100 ohms?

4AE-6.5 When Xc = 50 ohms, R = 100 ohms, and Xl = 25 ohms?

4AE-6.6 When Xc = 75 ohms, R = 100 ohms, and Xl = 50 ohms?

4AE-6.7 When Xc = 100 ohms, R = 100 ohms, and Xl = 75 ohms?

4AE-6.8 When Xc = 250 ohms, R = 1000 ohms, and Xl = 500 ohms?

4AE-6.9 When Xc = 50 ohms, R = 100 ohms, and Xl = 75 ohms?

4AE-6.10 When Xc = 100 ohms, R = 100 ohms, and Xl = 25 ohms?

4AE-7.1 Why would the true power (rate at which electrical energy is used) in a circuit be less than the product of the magnitudes of the alternating sine-wave voltage and the current?

4AE-7.2 In a circuit where an alternating sine-wave voltage and current are out of phase, how can the true power be determined?

4AE-7.3 What does the power factor equal, in an R-L circuit, with a phase angle of 60 degrees between the voltage and the current?

4AE-7.4 What does the power factor equal, in an R-L circuit, with a phase angle of 45 degrees between the voltage and the current?

4AE-7.5 What does the power factor equal, in an R-L circuit, with a phase angle of 30 degrees between the voltage and the current?

4AE-7.6 A circuit with a power factor of 0.2 has an input

voltage of 100 volts, and draws 4 amperes. How many watts are actually being consumed?

4AE-7.7 A circuit with a power factor of 0.6 has an input voltage of 200 volts, and draws 5 amperes. How many watts are actually being consumed?

4AE-8.1 What is the effective radiated power of a station in repeater operation with the following characteristics? 50 watts transmitter power output; 4 dB feedline loss; 3 dB duplexer and circulator loss; 6 dB antenna gain over a half-wave dipole.

4AE-8.2 50 watts transmitter power output; 5 dB feedline loss; 4 dB duplexer and circulator loss; 7 dB antenna gain.

4AE-8.3 75 watts transmitter power output; 4 dB feedline loss; 3 dB duplexer and circulator loss; 10 dB antenna gain.

4AE-8.4 75 watts transmitter power output; 5 dB feedline loss; 4 dB duplexer and circulator loss; 6 dB antenna gain.

4AE-8.5 100 watts transmitter power output; 4 dB feedline loss; 3 dB duplexer and circulator loss; 7 dB antenna gain.

4AE-8.6 100 watts transmitter power output; 5 dB feedline loss; 4 dB duplexer and circulator loss; 10 dB antenna gain.

4AE-8.7 120 watts transmitter power output; 5 dB feedline loss; 4 dB duplexer and circulator loss; 6 dB antenna gain.

4AE-8.8 150 watts transmitter power output; 4 dB feedline loss; 3 dB duplexer and circulator loss; 7 dB antenna gain.

4AE-8.9 200 watts transmitter power output; 4 dB feedline loss; 4 dB duplexer and circulator loss; 10 dB antenna gain.

4AE-8.10 200 watts transmitter power output; 4 dB feedline loss; 3 dB duplexer and circulator loss; 6 dB antenna gain.

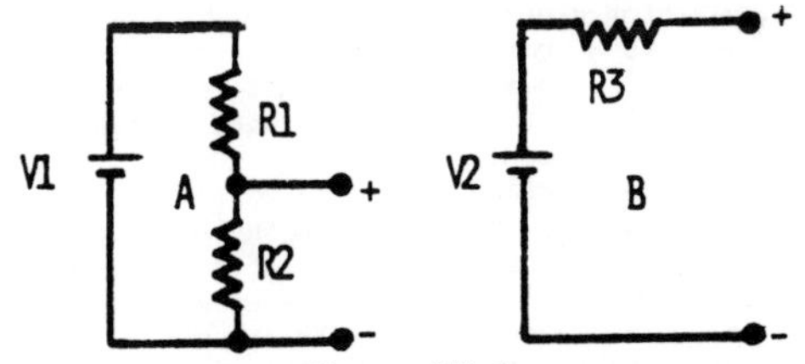

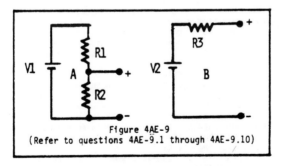

Figure 4AE-9
(Refer to questions 4AE-9.1 through 4AE-9.10)

4AE-9.1 To replace circuit A with circuit B in Figure 4AE-9, what voltage V2 and resistance R3 would be required to obtain exactly the same voltage and current characteristics, when V1 = 8 volts; R1 = 8K ohms; and R2 = 8K ohms?

4AE-9.2 When V1 = 8 volts; R1 = 16K ohms; and R2 = 8K ohms?

4AE-9.3 When V1 = 8 volts; R1 = 8K ohms; and R2 = 16K ohms?

4AE-9.4 When V1 = 10 volts; R1 = 10K ohms; and R2 = 10K ohms?

4AE-9.5 When V1 = 10 volts; R1 = 20K ohms; and R2 = 10K ohms?

4AE-9.6 When V1 = 10 volts; R1 = 10K ohms; and R2 = 20K ohms?

4AE-9.7 When V1 = 12 volts; R1 = 10K ohms; and R2 = 10K ohms?

4AE-9.8 When V1 = 12 volts; R1 = 20K ohms; ard R2 = 10K ohms?

4AE-9.9 When V1 = 12 volts; R1 = 10K ohms; and R2 = 20K ohms?

4AE-9.10 When V1 = 12 volts; R1 = 20K ohms; and R2 = 20K ohms?

## SUBELEMENT 4AF - Circuit Components (6 questions)

4AF-1.1 What is the schematic symbol for a semiconductor diode/rectifier? Label the anode and the cathode.

4AF-1.2 Structurally, what are the two main categories of semiconductor diodes?

4AF-1.3 What is the schematic symbol for a Zenner diode?

4AF-1.4 What are the two primary classifications of Zenner diodes?

4AF-1.5 What is the principal characteristic of a Zenner diode?

4AF-1.6 What is the range of voltage ratings available in Zenner diodes?

4AF-1.7 What is the schematic symbol for a tunnel diode?

4AF-1.8 What is the principal characteristic of a tunnel diode?

4AF-1.9 What special type of diode is capable of both amplification and oscillation?

4AF-1.10 What is the schematic symbol for a varactor diode?

4AF-1.11 What type of semiconductor diode varies its internal capacitance as the voltage applied to its terminals varies?

4AF-1.12 What is the principal characteristic of a varactor diode?

4AF-1.13 What is a common use of a varactor diode?

4AF-1.14 What is a common use of a hot-carrier diode?

4AF-1.15 What limits the maximum forward current in a junction diode?

4AF-1.16 How are junction diodes rated?

4AF-1.17 What is a common use for point-contact diodes?

4AF-1.18 What type of diode is made of a metal whisker touching a very small semi-conductor die?

4AF-1.19 What is common use for PIN diodes?

4AF-1.20 What special type of diode is often use for RF switches, attenuators, and various types of phase shifting devices?

4AF-2.1 What is the schematic symbol for a PNP transistor?

4AF-2.2 What is the schematic symbol for an NPN transistor?

4AF-2.3 What are the three terminals of a bipolar transistor?

4AF-2.4 What is the meaning of the term alpha with regard to bipolar transistor?

4AF-2.5 What is the term used to express the ratio of change in dc collector current to a change in emitter current in a bipolar transistor?

4AF-2.6 What is the meaning of the term beta with regard to bipolar transistors?

4AF-2.7 What is the term used to express the ratio of change in the dc collector current to a change in base current in a bipolar transistor?

4AF-2.8 What is the meaning of the term alpha cutoff frequency with regard to a bipolar transistor?

4AF-2.9 What is the term used to express that frequency at which the grounded base current gain has decreased to 0.7 of the gain obtainable at 1 kHz in a transistor?

4AF-2.10 What is the meaning of the term beta cutoff frequency with regard to a bipolar transistor?

4AF-2.11 What is the meaning of the term transition region with regard to a transistor?

4AF-2.12 What does it mean for a transistor to be fully saturated?

4AF-2.13 What does it mean for a transistor to be cut off?

4AF-2.14 What is the schematic symbol for a unijunction transistor?

4AF-2.15 What are the elements of a unijunction transistor?

4AF-2.16 For best efficiency and stability, where in the transistion region should a solid-state power amplifier be operated?

4AF-2.17 Name two elements widely used in semiconductor devices which exhibit both metalic and non-metalic characteristics.

4AF-3.1 What is the schematic symbol for a silicon controlled rectifier?

4AF-3.2 What are the three terminals of an SCR?

4AF-3.3 What are the two stable operating conditions of an SCR?

4AF-3.4 When an SCR is in the triggered or on condition, it has electrical characteristics similar to what other solid-state device (as measured between its cathode and anode)?

4AF-3.5 Under what operating condition does an SCR exhibit electrical characteristics similar to a foward-biased silicon rectifier?

4AF-3.6 What is the schematic symbol for a TRIAC?

4AF-3.7 What is the transistor called which is fabricated as two complementary SCRs in parallel with a common gate terminal?

4AF-3.8 What are the three terminals of a TRIAC?

4AF-4.0 What is the schematic symbol for a light-emitting diode?

4AF-4.1 What is the normal operating voltage and current for a light-emitting diode?

4AF-4.2 What type of bias is required for an LED to produce luminescense?

4AF-4.3 What are the advantages of using an LED?

4AF-4.4 What colors are available in LEDs?

4AF-4.5 What is the schematic symbol for a neon lamp?

4AF-4.6 What type neon lamp is usually used in amateur radio work?

4AF-4.7 What is the dc starting voltage for an NE-2 neon lamp?

4AF-4.8 What is the ac starting voltage for an NE-2 neon lamp?

4AF-4.9 How can a neon lamp be used to check for the presence of rf?

4AF-5.1 What would be the bandwidth of a good crystal lattice band-pass filter for SSB use?

4AF-5.2 What would be the bandwidth of a good crystal lattice band-pass filter for DSB voice use?

4AF-5.3 What is a crystal lattice filter?

4AF-5.4 What technique is often used by amateurs to construct low cost, high performance crystal lattice filters?

4AF-5.5 What determine the bandwidth and response shape in a crystal lattice filter?

SUBELEMENT 4AG - Practical Circuits (10 questions)

4AG-1.1 What is a linear electronic voltage regulator?

4AG-1.2 What is a switching electronic voltage regulator?

4AG-1.3 What device is usually used as a stable reference voltage in a linear voltage regulator?

4AG-1.4 What type of linear regulator is used in applications requiring efficient utilization of the primary power source?

4AG-1.5 What type of linear voltage regulator is used in applications where the load on the unregulated voltage source must be kept constant?

4AG-1.6 To obtain the best temperature stability, what should be the operating voltage of the reference diode in a linear voltage regulator?

4AG-1.7 What is the meaning of the term remote sensing with regard to a linear voltage regulator?

4AG-1.8 What is a three-terminal regulator?

4AG-1.9 What characteristics of a three-terminal regulator are important to consider?

4AG-2.1 What is the distinguishing feature of a Class A amplifier?

4AG-2.2 What class of amplifier is distinguished by the presence of output throughout the entire signal cycle and the input never goes into the cutoff region?

4AG-2.3 What is the distinguishing characteristic of a Class B amplifier?

4AG-2.4 What class of amplifier is distinguished by the flow of current in the output essentially in 180 degree pulses?

4AG-2.5 What is a Class AB amplifier?

4AG-2.6 What is the distinguishing feature of a Class C amplifier?

4AG-2.7 What class of amplifier is distinguished by the bias being set well beyond cutoff?

4AG-2.8 Which class of amplifier provides the highest efficiency?

4AG-2.9 Which class of amplifier has the highest linearity and least distortion?

4AG-2.10 Which class of amplifier has an operating angle of more than 180 degrees but less than 360 degrees when driven by a sine wave signal?

4AG-3.1 What is an L-network?

4AG-3.2 What is a pi-network?

4AG-3.3 What is a pi-L-network?

4AG-3.4 Does the L-, pi-, or pi-L-network provide the greatest harmonic suppression?

4AG-3.5 What are the three most commonly used networks to accomplish a match between an amplifying device and a transmission line?

4AG-3.6 Why are networks able to transform one impedance to another?

4AG-3.7 Which type of network offers the greater transformation ratio?

4AG-3.8 Why is the L-network of limited utility in impedance matching?

4AG-3.9 What is an advantage of using a pi-L-network instead of a pi-network for impedance matching between the final amplifier of a vacuum-tube type transmitter and a multiband antenna?

4AG-3.10 Which type of network provides the greatest harmonic suppression?

4AG-4.1 What are the three general groupings of filters?

4AG-4.2 What is constant-k filter?

4AG-4.3 What is an advantage of using a constant-k filter?

4AG-4.4 What is an m-derived filter?

4AG-4.5 What are the distinguishing features of a Butterworth filter?

4AG-4.6 What are the distinguishing features of a Chebyshev filter?

4AG-4.7 When would it be more desirable to use an m-derived filter over a constant-k filter?

4AG-5.1 What condition must exist for a circuit to oscillate?

4AG-5.2 What are the three major oscillator circuits used in amateur radio?

4AG-5.3 How is the positive feedback coupled to the input in a Hartley oscillator?

4AG-5.4 How is the positive feedback coupled to the input in a Colpitts oscillator?

4AG-5.5 How is the positive feedback coupled to the input in a Pierce oscillator?

4AG-5.6 Which of the three major oscillator circuits used in amateur radio utilizes a quartz crystal?

4AG-5.7 What is the piezoelectric effect?

4AG-5.8 What is the major advantage of a Pierce oscillator?

4AG-5.9 Which type of oscillator circuit is commonly used in a VFO?

4AG-5.10 Why is the Colpitts oscillator circuit commonly used in a VFO?

4AG-6.1 What is meant by the term modulation?

4AG-6.2 What are the two general categories of methods for producing frequency modulation?

4AG-6.3 What is a reactance modulator?

4AG-6.4 What is a balanced modulator?

4AG-6.5 How can a single sideband suppressed carrier signal be generated?

4AG-6.6 How can an AM signal be generated?

4AG-7.1 How is the efficiency of a power amplifier determined?

4AG-7.2 For reasonably efficient operation of a vacuum tube Class C amplifier, what should the plate-load resistance be with 1500 volts at the plate and 500 milliamperes plate current?

4AG-7.3 For reasonably efficient operation of a vacuum Class B amplifier, what should the load resistance be with 800 volts at the plate and 75 milliamperes plate current?

4AG-7.4 For reasonably efficient operation of a vacuum tube Class A operation what should the plate-load resistance be with 250 volts at the plate and 25 milliamperes plate current?

4AG-7.5 For reasonably efficient operation of a transistor amplifier, what should the load resistance be with 12 volts at the collector and 5 watts power output?

4AG-7.6 What is the flywheel effect?

4AG-7.7 How can a power amplifier be neutralized?

4AG-7.8 What order of Q is required by a tank-circuit sufficient to reduce harmonics to an acceptable level?

4AG-7.9 How can parasitic oscillations be eliminated from a power amplifier?

4AG-7.10 What is the procedure for tuning a power amplifier having an output pi-network?

4AG-8.1 What is the process of detection?

4AG-8.2 What is the principle of detection in a diode detector?

4AG-8.3 What is a product detector?

4AG-8.4 How are FM signals detected?

4AG-8.5 What is a frequency discriminator?

4AG-8.6 What is the mixing process?

4AG-8.7 What are the principal frequencies which appear at the output of a mixer circuit?

4AG-8.8 What are the advantages of the frequency-conversion process?

4AG-8.9 What occurs in a receiver when an excessive amount of signal energy reaches the mixer circuit?

4AG-9.1 How much gain should be used in the rf amplifier stage of a receiver?

4AG-9.2 Why should the rf amplifier stage of a receiver only have sufficient gain to allow weak signals to overcome noise generated in the first mixer stage?

4AG-9.3 What is the primary purpose of an rf amplifier in a receiver?

4AG-9.4 What is an i-f amplifier stage?

4AG-9.5 What factors should be considered when selecting an intermediate frequency?

4AG-9.6 What is the primary purpose of the first i-f amplifier stage in a receiver?

4AG-9.7 What is the primary purpose of the final i-f amplifier stage in a receiver?

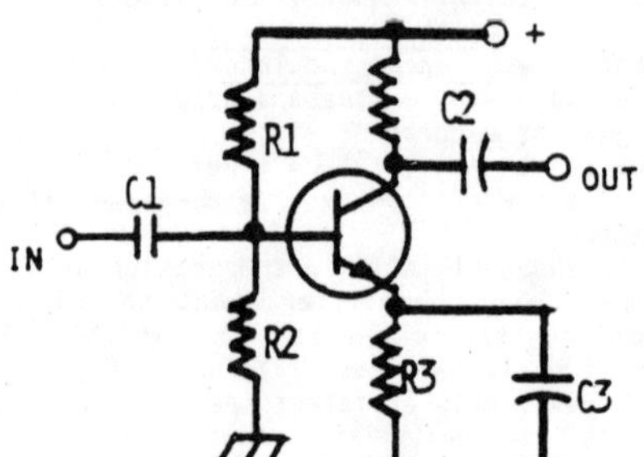

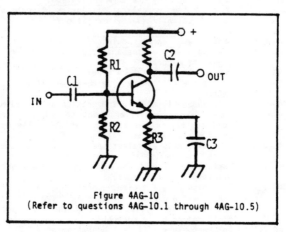

Figure 4AG-10
(Refer to questions 4AG-10.1 through 4AG-10.5)

4AG-10.1 What type of circuit is shown in Figure 4AG-10?

4AG-10.2 In Figure 4AG-10, what is the purpose of R1 and R2?

4AG-10.3 In Figure 4AG-10, what is the purpose of C1?
4AG-10.4 In Figure 4AG-10, what is the purpose of C3?
4AG-10.5 In Figure 4AG-10, what is the purpose of R3?

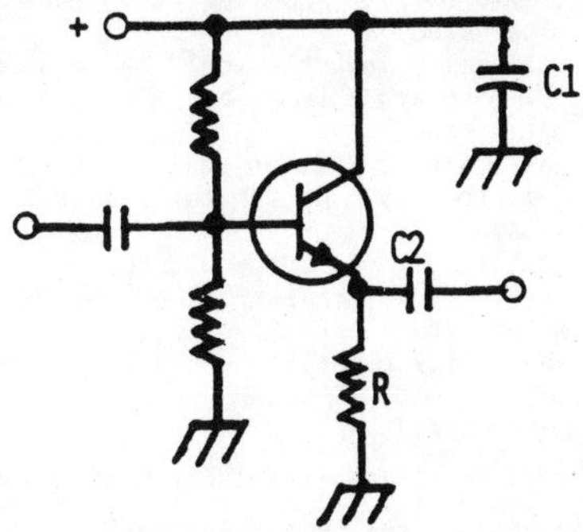

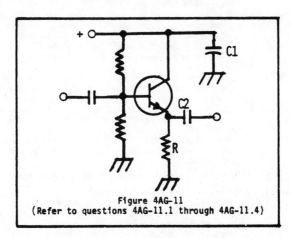

Figure 4AG-11
(Refer to questions 4AG-11.1 through 4AG-11.4)

4AG-11.1 What type of circuit is shown in Figure 4AG-11?
4AG-11.2 In Figure 4AG-11, what is the purpose of R?
4AG-11.3 In Figure 4AG-11, what is the purpose of C1?
4AG-11.4 In Figure 4AG-11, what is the purpose of C2?

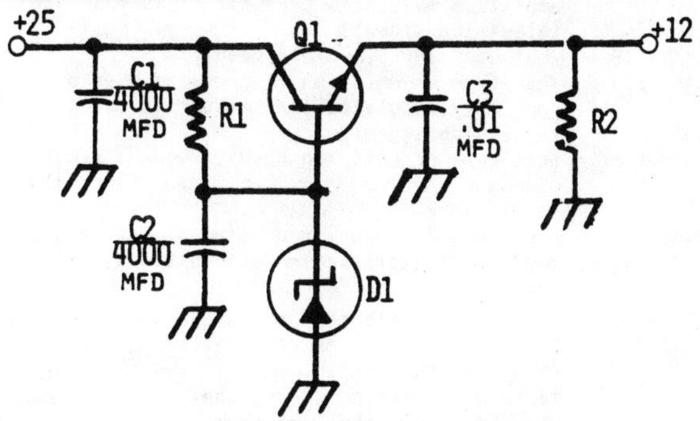


Figure 4AG-12
(Refer to questions 4AG-12.1 through 4AG-12.8)

4AG-12.1 What type of circuit is shown in Figure 4AG-12?
4AG-12.2 What is the purpose of D1 in the circuit shown in Figure 4AG-12?
4AG-12.3 What is the purpose of Q1 in the circuit shown in Figure 4AG-12?
4AG-12.4 What is the purpose of C1 in the circuit shown in Figure 4AG-12?
4AG-12.5 What is the purpose of C2 in the circuit shown in Figure 4AG-12?
4AG-12.6 What is the purpose of C3 in the circuit shown in

Figure 4AG-12?

4AG-12.7 What is the purpose of R1 in the circuit shown in Figure 4AG-12?

4AG-12.8 What is the purpose of R2 in the circuit shown in Figure 4AG-12?

4AG-13.1 What value capacitor would be required to tune a 20-microhenry inductor to resonate in the 80 meter band?

4AG-13.2 What value inductor would be required to tune a 100-picofarad capacitor to resonate in the 40 meter band?

4AG-13.3 What value capacitor would be required to tune a 2-microhenry inductor to resonate in the 20 meter band?

4AG-13.4 What value inductor would be required to tune a 15-picofarad capacitor to resonate in the 15 meter band?

4AG-13.5 What value capacitor would be required to tune a 100-microhenry inductor to resonate in the 160 meter band?

SUBELEMENT 4AH - Signals and Emissions (6 questions)

4AH-1.1 What does the symbol A4 mean?

4AH-1.2 What type of emission is produced when an amplitude modulated transmitter is modulated by a facsimile signal?

4AH-1.3 What is facsimile?

4AH-1.4 What does the symbol F4 mean?

4AH-1.5 What type of emission is produced when a frequency modulated transmitter is modulated by a facsimile signal?

4AH-1.6 What does the symbol A5 mean?

4AH-1.7 What type of emission is produced when an amplitude modulated transmitter is modulated by a television signal?

4AH-1.8 What does the symbol F5 mean?

4AH-1.9 What type of emission is produced when a frequency modulated transmitter is modulated by a television signal?

4AH-1.10 What type of emission results when a single sideband transmitter is used for slow-scan television?

4AH-2.1 How can a frequency modulated signal be produced?

4AH-2.2 How can an amplitude modulated signal be produced?

4AH-2.3 How can a single sideband signal be produced?

4AH-3.1 What is meant by the term deviation ratio?

4AH-3.2 In an FM transmitter, what is the term for the maximum deviation from the carrier frequency divided by the maximum audio modulating frequency?

4AH-3.3 What is the deviation ratio for an FM transmitter having a maximum frequency swing of plus or minus 5 kHz and accepting a maximum modulation rate of 3 kHz?

4AH-3.4 What is the deviation ratio for an FM transmitter having a maximum frequency swing of plus or minus 7.5 kHz and accepting a maximum modulation rate of 3.5 kHz?

4AH-4.1 What is meant by the term modulation index?

4AH-4.2 In an FM transmitter, what is the term for the ratio between the deviation of a frequency modulated signal and the modulating frequency?

4AH-4.3 How does the modulation index of a phase-modulated emission vary with the modulated frequency?

4AH-4.4 In an FM transmitter having a maximum frequency deviation of 3000 Hz either side of the carrier frequency, what is the modulation index when the modulating frequency is 1000 Hz?

4AH-4.5 What is the modulation index of a frequency modulated transmitter producing an instantaneous carrier deviation of 6 kHz when modulated with a 2 kHz modulating frequency?

4AH-5.1 What are electromagnetic waves?

4AH-5.2 What is a wave front?

4AH-5.3 At what speed do electromagnetic waves travel in free space?

4AH-5.4 What are the two interrelated fields considered to make up an electromagnetic wave?

4AH-5.5 Why cannot electromagnetic waves penetrate a good conductor to any great extent?

4AH-6.1 What is meant by referring to electromagnetic waves travel in free space?

4AH-6.2 What is meant by referring to electromagnetic waves as horizontally polarized?

4AH-6.3 What is meant by referring to electromagnetic waves as having circular polarization?

4AH-6.4 When the electric field is perpendicular to the surface of the earth, what is the polarization of the electromagnetic wave?

4AH-6.5 When the magnetic field is parallel to the surface of the earth, what is the polarization of the electromagnetic wave?

4AH-6.6 When the magnetic field is perpendicular to the surface of the earth, what is the polarization of the electromagnetic field?

4AH-6.7 When the electric field is parallel to the surface of the earth, what is the polarization of the electromagnetic wave?

4AH-7.1 What is a sine wave?

4AH-7.2 How many times does a sine wave cross the zero axis in one complete cycle?

4AH-7.3 How many degrees are there in one complete sine wave cycle?

4AH-7.4 What is the period of a wave?

4AH-7.5 What is a square wave?

4AH-7.6 What is a wave called which abruptly changes back and forth between two voltage levels and which remains an equal time at each level?

4AH-7.7 Which sine waves make up a square wave?

4AH-7.8 What type of wave is made up of sine waves of the fundamental frequency and all the odd harmonics?

4AH-7.9 What is a sawtooth wave?

4AH-7.10 What type of wave is characterized by a rise time significantly faster than the fall time (or vice versa)?

4AH-7.11 Which sine waves make up a sawtooth wave?

4AH-7.12 What type of wave is made up of sine waves at the fundamental frequency and all the harmonics?

4AH-8.1 What is the meaning of the term root mean square value of an ac voltage?

4AH-8.2 What is the term used in reference to a dc voltage that would cause the same heating in a resistor as a certain value of ac voltage?

4AH-8.3 What would be the most accurate way of determining the rms voltage of a complex waveform?

4AH-8.4 What is the rms voltage at a common household electrical power outlet?
4AH-8.5 What is the peak voltage at a common household electrical outlet?
4AH-8.6 What is the peak-to-peak voltage at a common household electrical outlet?
4AH-8.7 What is the rms voltage of a 165 volt peak pure sine wave?
4AH-8.8 What is the rms voltage of a 331 volt peak-to-peak pure sine wave?
4AH-9.1 For many types of voices, what is the ratio of PEP to average power during a modulation peak in a single sideband transmission?
4AH-9.2 With a single sideband, suppressed carrier A3 emission transmission, what determines the PEP-to-average power ratio?
4AH-10.1 Where is the noise generated which primarily determines the signal-to-noise ratio in a radio receiver for the 160 meter band?
4AH-10.2 Where is the noise generated which primarily determines the signal-to-noise ratio in a radio receiver for the 2 meter band?
4AH-10.3 Where is the noise generated which primarily determines the signal-to-noise ratio in a radio receiver for the 220-225 MHz band?
4AH-10.4 Where is the noise generated which primarily determines the signal-to-noise ratio in a radio receiver for the 420-450 MHz band?

SUBELEMENT 4AI - Antenna and Feedlines (5 questions)

4AI-1.1 What is meant by the term antenna gain?
4AI-1.2 What is the term for a numerical ratio which relates the performance of one antenna to that of another real or theoretical antenna?
4AI-1.3 What is meant by the term antenna beamwidth?
4AI-1.4 How do you find the beamwidth of a rotatable beam antenna?
4AI-2.1 What is a trap antenna?
4AI-2.2 What is an advantage of using a trap antanna?
4AI-2.3 What is a disadvantage of using a trap antenna?
4AI-2.4 What is the principle of a trap antenna?
4AI-3.1 What is a parasitic element of an antenna?
4AI-3.2 How does a parasitic element generate an electromagnetic field?
4AI-3.3 For a parasitic element beam antenna, how does the length of the reflector compare with that of the driven element?
4AI-3.4 For a parasitic element beam antenna, how does the length of the director compare with that of the driven element?
4AI-4.1 What is meant by the term radiation resistance for an antenna?
4AI-4.2 What is the term used for an equivalent resistance which would dissipate the same amount of energy as that radiated from an antenna?
4AI-4.3 Why is the value of the radiation resistance of an antenna important?
4AI-4.4 What are the factors that determine the radiation resistance of an antenna?
4AI-5.1 What is a driven element of an antenna?
4AI-5.2 What is the usual electrical length of a driven element in an amateur beam for HF?
4AI-5.3 What is the term for an antenna element which

is supplied power from a transmitter through a transmission line?

4AI-6.1 What is meant by the term antenna efficiency?

4AI-6.2 What is the term for the ratio of the radiation resistance of an antenna to the total resistance of the system?

4AI-6.3 What is included in the total resistance of an antenna system?

4AI-6.4 How can the antenna efficiency of a HF grounded vertical antenna be made comparable to that of a half-wave antenna?

4AI-6.5 Why does a half-wave antenna operate at very high efficiency?

4AI-7.1 What is a folded dipole antenna?

4AI-7.2 How does the bandwidth of a folded dipole antenna compare with that of a simple dipole antenna?

4AI-7.3 What is the characteristic impedance at the center of a folded dipole antenna?

4AI-7.4 What is a tripole antenna?

4AI-7.5 What is the characteristic impedance of a tripole antenna?

4AI-8.1 What is the meaning of the term velocity factor of a transmission line?

4AI-8.2 What is the term for the ratio of actual velocity at which a signal travels through a line to the speed of light in a vacuum?

4AI-8.3 What is the velocity factor for a typical coaxial cable?

4AI-8.4 What determines the velocity factor in a transmission line?

4AI-8.5 Why is the physical length of a coaxial cable transmission line shorter than its electrical length?

4AI-9.1 What would be the physical length of a coaxial transmission line which is electrically one-quarter wave length long at 14.1 MHz? (assume a velocity factor of 0.66.)

4AI-9.2 At a frequency of 7.2 MHz, what is the physical length of a coaxial transmission line with an electrical length of one-quarter wavelength and a velocity factor of 0.66?

4AI-9.3 What is the physical length of a parallel antenna feedline which is electrically one-half wavelength long at 14.10 MHz? (assume a velocity factor of 0.82.)

4AI-9.4 What is the physical length of a twin lead transmission feedline at 3.65 MHz? (assume a velocity factor of 0.80.)

4AI-10.1 In a half-wave antenna, where do the current nodes exist?

4AI-10.2 In a half-wave antenna, where do the voltage nodes exist?

4AI-10.3 At the ends of a half-wave antenna, what values of current and voltage exist compared to the remainder of the antenna?

4AI-10.4 At the center of a half-wave antenna, what values of voltage and current exist compared to the remainder of the antenna?

4AI-11.1 Why is the inductance required for a base loaded HF mobile antenna less than that for an inductance placed further up the whip?

4AI-11.2 What happens to the base feed point of a fixed length HF mobile antenna as the frequency of operation is lowered?

4AI-11.3 Why should an HF mobile antenna loading coil have a high ratio of reactance to resistance?

4AI-11.4 Why is a loading coil used with an HF mobile antenna?

4AI-12.1 For a shortened vertical antenna, where should a loading coil be placed to minimize losses and produce the most effective performance?

4AI-12.2 What happens to the bandwidth of an antenna as it is shortened through the use of loading coils?

4AI-12.3 Why are self-resonant antennas popular in amateur radio?

4AI-12.4 What is an advantage of using top loading in a shortened HF vertical antenna?

# QUESTIONS FOR THE ELEMENT 4B AMATEUR RADIO OPERATOR LICENSE EXAMINATION

The questions used by the examiner in an Element 4B written examination must be taken from the following list. CANDIDATES SHOULD NOT TRY TO MEMORIZE THESE QUESTIONS! The examination is a test of the knowledge a person needs in order to operate an amateur radio station properly. It is not a test of a person's ability to answer questions by rote. See PR Bulletin 1035, Study Guide For The Amateur Radio Operator Examinations for more information.

The questions must be used exactly as stated in this Bulletin. The Volunteer-Examiner Coordinators may use their discretion as to the form of the examination answers. They may be essay, multiple-choice or single answer type. See Secticn 97.27 of the Commission's Rules.

Direct any inquiries concerning these questions to the Volunteer-Examiner Coordinator. Do not contact the FCC.

SUBELEMENT 4BA - Rules and Regulations (8 questions)

4BA-1.1 What exclusive frequency privileges in the 75/80 meter band are authorized to Amateur Extra class operators?

4BA-1.2 What exclusive frequency privileges in the 40 meter band are authorized to Amateur Extra class operators?

4BA-1.3 What exclusive frequency privileges in the 20 meter band are authorized to Amateur Extra class operators?

4BA-1.4 What exclusive frequency privileges in the 15 meter band are authorized to Amateur Extra class operators?

4BA-1.5 What frequencies are available to amateur radio operators in the 75 meter band in I.T.U. Region 3?

4BA-1.6 What frequencies are available to amateur radio operators in the 40 meter band in I.T.U. Region 3?

4BA-1.7 What are the purposes of the Amateur-Satellite Service?

4BA-2.1 Which amateur radio stations are eligible for space operation?

4BA-2.2 Which amateur radio stations are eligible for earth operation?

4BA-2.3 Which amateur radio stations are eligible for telecommand operation?

4BA-2.4 What are the requirements for space operation?

4BA-2.5 What frequencies are availab'e for space operation?

4BA-2.6 In what time period must the licensee of a station in space operation give written pre-space notification to the FCC?

4BA-2.7 In what time period is the licensee of a station in space operation required to give written in-space notification to the FCC?

4BA-2.8 In what time period is the licesee of a station in space operation required to give written post-space notification to the FCC?

4BA-3.1 How much must the mean power of any spurious

emission or radiation from an amateur transmitter be attenuated when the carrier frequency is below 30 MHz?

4BA-3.2 How much must the mean power of any spurious emission or radiation from an amateur transmitter be attenuated when the carrier frequency is above 30 MHz?

4BA-3.3 What is a spurious emission or radiation?

4BA-3.4 What can the FCC require the licensee to do if any spurious radiation causes harmful interference to the reception of another radio station?

4BA-4.1 What are the special conditions an amateur radio station operated aboard a ship or aircraft must comply with?

4BA-4.2 Is an FCC amateur license or reciprocal permit required in order to operate an amateur radio station aboard a vessel registered in the United States?

4BA-5.1 What is RACES?

4BA-5.2 What is the purpose of RACES?

4BA-5.3 Which amateur radio stations may be operated in RACES?

4BA-5.4 Which persons may be control operators of RACES stations?

4BA-5.5 What are the points of communications for amateur stations operated in RACES and certified by the responsible civil defense organization as registered with that organization?

4BA-5.6 What are permissible communications in RACES?

4BA-6.1 What are the points of communications for an amateur radio station?

4BA-7.1 Is a person who is not a citizen of the United States eligible to obtain an amateur radio license from the FCC?

4BA-7.2 What is a Reciprocal Operating Permit?

4BA-7.3 Who is eligible for a Reciprocal Operating Permit?

4BA-7.4 Is a person who is a citizen of the United States eligible to obtain a Reciprocal Operating Permit?

4BA-7.5 Is a person eligible for a Reciprocal Operating Permit if that person is not a citizen of the country issuing the foreign amateur radio license?

4BA-7.6 What are the operator frequency privileges authorized by a Reciprocal Operating Permit?

4BA-7.7 How does the alien amateur identify the station when operating under a Reciprocal Operating Permit?

4BA-8.1 What is a Volunteer-Examiner Coordinator (VEC)?

4BA-8.2 What are the qualifications to be a VEC?

4BA-8.3 What are the functions of a VEC?

4BA-9.1 What is Examination Element 1(A)?

4BA-9.2 What is Examination Element 1(B)?

4BA-9.3 What is Examination Element 1(C)?

4BA-9.4 What is Examination Element 2?

4BA-9.5 What is Examination Element 3?

4BA-9.6 What is Examination Element 4A?

4BA-9.7 What is Examination Element 4B?

4BA-10.1 What examination elements are required for an Amateur Extra class operator license?

4BA-10.2 What examination elements are required for an Advanced class operator license?

4BA-10.3 What examination elements are required for a

General class operator license?

4BA-10.4 What examination elements are required for a Technician class operator license?

4BA-11.1 What examination credit must be given to an applicant who holds a valid Novice operator class license?

4BA-11.2 What examination credit must be given to an applicant who holds a valid Technician operator class license?

4BA-11.3 What examination credit must be given to an applicant who holds a valid General operator class license?

4BA-11.4 What examination credit must be given to an applicant who holds a valid Advanced operator class license?

4BA-11.5 What examination credit, if any, may be given to an applicant who holds a valid amateur radio operator license issued by another country?

4BA-11.6 What examination credit, if any, may be given to an applicant who holds a valid amateur radio operator issued by any other United States government agency than the FCC?

4BA-11.7 What examination credit must be given to an applicant who holds a valid FCC commercial radiotelegraph license?

4BA-11.8 What examination credit must be given to the holder of a valid Certificate of Successful Completion of Examination?

4BA-12.1 Who determines where and when examinations for amateur operator licenses are to be administered?

4BA-12.2 Where must the examiner(s) be and what must the examiner(s) be doing during an examination?

4BA-12.3 Who is responsible for the proper conduct and necessary supervision during an examination?

4BA-12.4 What should an examiner do upon failure of the candidate to comply with the examiner's instructions?

4BA-12.5 What must the candidate do at the completion of the examination?

4BA-12.6 How long must an applicant wait to retake an examination element they have failed?

4BA-13.1 Who must prepare Examination Element 1(B)?

4BA-13.2 Who must prepare Examination Element 1(C)?

4BA-13.3 Who must prepare Examination Element 3?

4BA-13.4 Who must prepare Examination Element 4A?

4BA-13.5 Who must prepare Examination Element 4B?

4BA-13.6 Where are the questions listed that must be used a written examinations?

4BA-14.1 When must the test papers be graded?

4BA-14.2 Who must grade the test papers?

4BA-14.3 How does the examiner(s) inform a candidate who does not score a passing grade?

4BA-14.4 What must the examiner(s) do when the candidate scores a passing grade?

4BA-14.5 How long after the administration of a successful examination for a Technician, General, Advanced or Amateur Extra class operator license must the examiners submit the candidate's applications and test papers?

4BA-14.6 To whom do the examiners submit successful candidates' applications and test papers?

4BA-15.1 What are the requirements for a volunteer examiner administering an examination for Technician, General, Advanced, or Amateur Extra class operator?

4BA-16.1 Under what authority may a successful applicant operate his/her amateur radio station with the rights and privileges of the higher operator class for which the applicant has passed the appropriate examinations?

4BA-16.2 What is a Certificate of Successful Completion of Examination?

4BA-16.3 How long may a successful applicant operate his/her station under Section 97.35 with the rights and privileges of the higher operator class for which the applicant has passed the appropriate examinations?

4BA-16.4 How must the station call sign be amended when operating under the temporary authority authorized by Section 97.35?

SUBELEMENT 4BB - Operating Procedures (2 questions)

4BB-1.1 Why does the downlink frequency appear to vary by several kHz during a low earth orbit amateur satellite pass?

4BB-1.2 What are the two basic types of linear transponders used in amateur satellites?

4BB-1.3 What is a linear transponder?

4BB-1.4 What is Mode A in an amateur satellite?

4BB-1.5 What is Mode B in an amateur satellite?

4BB-1.6 What is Mode J in an amateur satellite?

4BB-1.7 What is Mode L in an amateur satellite?

4BB-1.8 What is a ascending pass for an amateur satellite?

4BB-1.9 What is a decending pass for an amateur satellite?

4BB-1.10 What is the period of an amateur satellite?

4BB-2.1 How often is a new frame transmitted in a fast-scan television system?

4BB-2.2 How many horizontal lines make up a fast-scan television frame?

4BB-2.3 How is the interlace scanning pattern generated in a fast-scan television system?

4BB-2.4 What is blanking in a video signal?

4BB-2.5 What is the standard video voltage level between the sync tip and the whitest white at TV camera outputs and modulator inputs?

4BB-2.6 How is audio transmitted in an amateur television system?

4BB-2.7 What is the bandwidth of a fast-scan television transmission?

4BB-2.8 What is the standard video level, in percent PEP, for black?

4BB-2.9 What is the standard video level, in percent PEP, for white?

4BB-2.10 What is the standard video level, in percent PEP, for blanking?

SUBELEMENT 4BC - Radio Wave Propagation (1 Question)

4BC-1.1 How far apart may stations communicating by moonbounce be separated?

4BC-1.2 What characterizes libration fading of an E-M-E signal?

4BC-1.3 What are the best days to schedule E-M-E contacts?

4BC-1.4 What type of receiving system is required for E-M-E communications?

| | |
|---|---|
| 4BC-1.5 | What type of transmitting system is required for E-M-E communications? |
| 4BC-2.1 | When the earth's atmosphere is struck by a meteor, a cylindrical region of free electrons is formed at what level of the ionosphere? |
| 4BC-2.2 | Which range of frequencies are best suited for meteor-burst transmissions? |
| 4BC-3.1 | What is trans-equatorial propagation? |
| 4BC-3.2 | What is the maximum range for signals using trans-equatorial propagation? |
| 4BC-3.3 | What is the best time of day for trans-equatorial propagation? |

SUBELEMENT 4BD - Amateur Radio Practice (3 Questions)

| | |
|---|---|
| 4BD-1.1 | What is a spectrum analyzer? |
| 4BD-1.2 | What type of instrument may be used to observe electrical signals in the frequency domain? |
| 4BD-1.3 | How does a spectrum analyzer differ from a conventional oscilloscope? |
| 4BD-1.4 | What does the horizontal axis of a spectrum analyzer display? |
| 4BD-1.5 | What does the vertical axis of a spectrum analyzer display? |
| 4BD-2.1 | What is a logic probe? |
| 4BD-2.2 | How is a logic probe used? |
| 4BD-2.3 | What does a logic probe indicate? |
| 4BD-2.4 | Where is a logic probe used? |
| 4BD-2.5 | What advantage does a logic probe have over a voltmeter in monitoring logic states in a circuit? |
| 4BD-3.1 | What is one of the most significant deterrents to effective signal reception at a mobile station? |
| 4BD-3.2 | What is the proper procedure for suppressing electrical noise in a mobile station? |
| 4BD-3.3 | How can ferrite beads be used to suppress ignition noise in newer vehicles? |
| 4BD-3.4 | How can metal bonding of the vehicle reduce the level of spark plug noise? |
| 4BD-3.5 | What are the main areas of a vehicle to bond in order to reduce electrical noise? |
| 4BD-3.6 | How can alternator whine be minimized? |
| 4BD-3.7 | How can conducted noise in a vehicle be minimized? |
| 4BD-3.8 | How can the electrical noise generated by automobile instruments be suppressed? |
| 4BD-3.9 | How can conducted and radiated noise caused by an automobile alternator be suppressed? |
| 4BD-3.10 | What can cause corona-discharge noise? |
| 4BD-4.1 | What is the main drawback of a wire-loop antenna for direction finding? |
| 4BD-4.2 | What antenna characteristics are desirable for direction finding? |
| 4BD-4.3 | What is the triangulation method of direction finding? |
| 4BD-4.4 | Why is an RF attenuator desirable in a receiver used for direction finding? |
| 4BD-4.5 | What is a sense antenna? |
| 4BD-4.6 | What is an Adcock antenna? |
| 4BD-4.7 | What is a loop antenna? |
| 4BD-4.8 | How can the output voltage of a loop antenna be increased? |
| 4BD-4.9 | What is the night effect sometimes experienced in direction finding? |
| 4BD-4.10 | What type of terrain can cause errors in direction finding? |

SUBELEMENT 4BE - Electrical Principles (7 Questions)

4BE-1.1 What is the photoconductive effect?

4BE-1.2 What happens to the conductivity of a photoconductive material when light shines upon it?

4BE-1.3 What happens to the resistance of a photoconductive material when light shines upon it?

4BE-1.4 What happens to the conductivity of a semiconductor junction when it is illiminated?

4BE-1.5 What is an optocoupler?

4BE-1.6 What is an optoisolator?

4BE-1.7 What is an optical shaft-encoder?

4BE-1.8 What is a solid-state relay?

4BE-1.9 What is one of the cleanest ways to mate solid-state circuits operating at widely deferring voltages?

4BE-1.10 What does the photoconductive effect in crystalline solids produce a noticable change in?

4BE-2.1 What is the meaning of the term time constant of an RC circuit?

4BE-2.2 What is the meaning of the term time constant of an RL circuit?

4BE-2.3 What is the term for the time required for the capacitor in an RC circuit to be charged to 63.2% of the supply voltage?

4BE-2.4 What is the term for the time required for the inductor in a RL circuit to build to 63.2% of the maximum value?

4BE-2.5 What is the term for the time it takes for a charged capacitor in an RC circuit to discharge to 36.8% of its initial value?

4BE-2.6 What is meant by back emf?

4BE-2.7 How is the time constant for an RC circuit expressed?

4BE-2.8 How is the time constant for an RL circuit expressed?

4BE-2.9 To what percentage of the supply voltage will a capacitor in an RC circuit be charged after two time constants?

4BE-2.10 To what percentage of the starting voltage will a capacitor in an RC circuit be discharged after two time constants.

4BE-3.1 What is the time constant of a circuit having a 100-microfarad capacitor in series with a 470-kohm resistor?

4BE-3.2 What is the time constant of a circuit having a 220-microfarad capacitor in parallel with a 1-megohm resistor?

4BE-3.3 What is the time constant of a circuit having two 100-microfarad capacitors and two 470-K ohm resistors all in series?

4BE-3.4 What is the time constant of a circuit having two 100-microfarad capacitors and two 470-K ohms resistors all in parallel?

4BE-3.5 What is the time constant of a circuit having two 220-microfarad capacitors and two 1-megohm resistors all in series?

4BE-3.6 What is the time constant of a circuit having two 220-microfarad capacitors and two 1-megohm resistors all in parallel?

4BE-3.7 What is the time constant of a circuit having one 100-microfarad, one 220-microfarad capacitor and one 470-kohm and one 1-megohm resistor all in series?

4BE-3.8 What is the time constant of a circuit having a 470-microfarad capacitor and a 1-megohm resistor in parallel?

4BE-3.9 What is the time constant of a circuit having a 470-microfarad capacitor in series with a 470-kohm resistor?

4BE-3.10 What is the time constant of a circuit having a 220-microfarad capacitor in series with a 470-kohm resistor?

4BE-3.11 A 0.01-microfarad capacitor has a 2-megohm resistor connected across it. If the capacitor is initially charged to 20-volts, how long will it take for the voltage to die down to 7.36-volts?

4BE-3.12 A 0.01 microfarad capacitor has a 2-megohm resistor connected across it. If the capacitor is initially charged to 20-volts, how long will it take for the voltage to die down to 2.71-volts?

4BE-3.13 A 0.01 microfarad capacitor has a 2-megohm resistor connected across it. If the capacitor is initially charged to 20-volts, how long will it take for the voltage to die down to 1-volt?

4BE-3.14 A 0.01 microfarad capacitor has 2-megohm resistor connected across it. If the capacitor is initially charged to 20-volts, how long will it take for the voltage to die down to 0.37-volts?

4BE-3.15 A 0.01 microfarad capacitor has a 2-megohm resistor connected across it. If the capacitor is initially charged to 20-volts, how long will it take for the voltage to die down to 0.13-volts?

4BE-3.16 A 450-microfarad capacitor has a 1-megohm resistor connected across it. If the capacitor is initially charged to 800-volts, how long will it take for the voltage to die down to 294-volts?

4BE-3.17 A 450-microfarad capacitor has a 1-megohm resistor connected across it. If the capacitor is initially charged to 800-volts, how long will it take for the voltage to die down to 108-volts?

4BE-3.18 A 450-microfarad capacitor has a 1-megohm resistor connected across it. If the capacitor is initially charged to 800-volts, how long will it take for the voltage to die down to 39.9-volts?

4BE-3.19 A 450-microfarad capacitor has a 1-megohm resistor connected across it. If the capacitor is initially charged to 800-volts, how long will it take for the voltage to die down to 40.2-volts?

4BE-3.20 A 450-microfarad capacitor has a 1-megohm resistor connected across it. If the capacitor is initially charged to 800-volts, how long will it take for the voltage to die down to 14.8-volts?

4BE-4.1 What is a Smith Chart?

4BE-4.2 What type of coordinate system is used in a Smith Chart?

4BE-4.3 What type of problems can be solved in a Smith Chart?

4BE-4.4 What are the two families of circles which make up a Smith Chart?

4BE-4.5 What is the only straight line on a Smith Chart?

4BE-4.6 What is the process of normalizing with regard to a Smith Chart?

4BE-4.7 What are the curved lines on a Smith Chart?

4BE-4.8 What is the third family of circles which are added to a Smith Chart during the process of solving problems?

4BE-4.9 What are the four basic steps for performing calculations on a Smith Chart?

4BE-4.10 How are the wavelength scales on a Smith Chart calibrated?

4BE-5.1 What is the impedance of a network comprised of a 0.1-microhenry inductor in series with 20-ohms resistance, at 30 MHz?

4BE-5.2 What is the impedance of a network comprised of a 0.1-microhenry inductor in series with 30-ohms resistance, at 5 MHz?

4BE-5.3 What is the impedance of a network comprised of a 10-microhenry inductor in series with 40-ohms resistance, at 500 MHz?

4BE-5.4 What is the impedance of a network comprised of a 1.0-millihenry inductor in series with 200-ohms resistance, at 30 kHz?

4BE-5.5 What is the impedance of a network comprised of a 10-millihenry inductor in series with 600-ohms resistance, at 10 kHz?

4BE-5.6 What is the impedance of a network comprised of a 100-picofarad capacitor in parallel with 4000-ohms resistance, at 500 kHz?

4BE-5.7 What is the impedance of a network comprised of a 0.001-microfarad capacitor in series with 400-ohms resistance, at 500 kHz?

4BE-5.8 What is the impedance of a network comprised of a 0.01-microfarad capacitor in parallel with 300-ohms resistance, at 50 kHz?

4BE-5.9 What is the impedance of a network comprised of a 0.1-microfarad capacitor in series with 40-ohms resistance, at 50 kHz?

4BE-5.10 What is the impedance of a network comprised of a 1.0-microfarad capacitor in parallel with 30-ohms resistance, at 5 MHz?

4BE-6.1 What is the impedance and phase angle of a network comprised of a 100-ohm reactance inductor in series with 100-ohms resistance?

4BE-6.2 What is the impedance and phase angle of a network comprised of a 100-ohm reactance inductor, 100 ohms capacitive reactance, and 100-ohms resistance, all in series?

4BE-6.3 What is the impedance and phase angle of a network comprised of a 100-ohm reactance capicitor in series with 100-ohms resistance?

4BE-6.4 What is the impedance and phase angle of a network comprised of a 100-ohm reactance capacitor in parallel with 100-ohms resistance?

4BE-6.5 What is the impedance and phase angle of a network comprised of a 300-ohm reactance inductor in series with 400-ohms resistance?

4BE-6.6 What is the impedance and phase angle of a network comprised of a 400-ohm reactance capacitor in series with 300-ohms resistance?

4BE-6.7 What is the impedance and phase angle of a network comprised of a 300-ohm reactance capacitor, 600-ohms reactance inductor, and 400-ohms resistance, all in series?

4BE-6.8 What is the impedance and phase angle of a network comprised of a 400-ohm reactance inductor in series with 300-ohms resistance?

4BE-6.9 What is the impedance and phase angle of a network comprised of a 100-ohm reactance inductor in parallel with 100-ohms resistance?

4BE-6.10 What is the impedance and phase angle of a network comprised of a 300-ohm reactance capacitor in series with 400-ohms resistance?

SUBELEMENT 4BF - Circuit Components (5 questions)

4BF-1.1 What is the schematic symbol for an n-channel junction FET?

4BF-1.2 What is the schematic symbol for a p-channel junction FET?

4BF-1.3 How does the input impedance of a field-effect transistor compare with that of a bipolar transistor?

4BF-1.4 What are the two basic types of field-effect transistors?

4BF-1.5 What is the schematic symbol for an n-channel MOSFET?

4BF-1.6 What is the schematic symbol for a p-channel MOSFET?

4BF-1.7 What are the three terminals of a field-effect transistor?

4BF-1.8 What is the schematic symbol for an n-channel dual-gate MOSFET?

4BF-1.9 What is the schematic symbol for a p-channel dual-gate MOSFET?

4BF-1.10 How do the three terminals of a FET practically equate to the terminals of a triode vacuum tube?

4BF-1.11 What is the input impedance of a typical field-effect transistor?

4BF-1.12 What is a depletion-mode FET?

4BF-1.13 What is an enhancement-mode FET?

4BF-1.14 What is a CMOS IC?

4BF-1.15 Why are special precautions necessary in handling FET and CMOS devices?

4BF-1.16 Why do many MOSFET devices have built-in gate-protective Zener diodes?

4BF-2.1 What is the schematic symbol for an operational amplifier?

4BF-2.2 What is an operational amplifier?

4BF-2.3 What would be the characteristics of the ideal op-amp?

4BF-2.4 What determines the gain of an op-amp circuit?

4BF-2.5 What is an inverting op-amp?

4BF-2.6 What is a non-inverting op-amp?

4BF-2.7 What is a difference op-amp?

4BF-2.8 What is a phase-locked loop?

4BF-2.9 What functions are performed by a phase-locked loop IC?

4BF-2.10 How does a PLL function?

4BF-3.1 What is the recommended power supply voltage for 7400 series integrated circuits?

4BF-3.2 What type of digital-logic is used in the 7400 series integrated circuits?

4BF-3.3 If TTL device inputs are left open, what logic state do they assume?

4BF-3.4 When operated with a plus 5-volt power supply, what level of input voltage is defined as "high" in a TTL device?

4BF-3.5 When operated with a plus 5-volt power supply, what level of input voltage is defined as "low" in a TTL device?

4BF-3.6 Why should circuits containing TTL devices have serveral bypass capacitors per pc board?

4BF-3.7 What is an AND gate?

4BF-3.8 What is an OR gate?
4BF-3.9 What is a NOT gate?
4BF-3.10 What is a NOR gate?
4BF-3.11 What is a NAND gate?
4BF-3.12 What is a 7404 integrated circuit?
4BF-3.13 What is a 7447 integrated circuit?
4BF-3.14 What is a 7493 integrated circuit?
4BF-3.15 What is a truth table?
4BF-4.1 What does the term CMOS mean?
4BF-4.2 What type of integrated circuit is the 4000 series?
4BF-4.3 What is one major advantage of CMOS over other devices?
4BF-4.5 Why do 4000 series integrated circuits have high immunity to noise on the input signal or power supply?
4BF-5.1 What is a vidicon?
4BF-5.2 How is the electron beam deflected in a vidicon?
4BF-5.3 What is the output impedance of a vidicon?
4BF-5.4 What type of CRT deflection is better when high frequency waves are to be displ_yed on the screen?
4BF-5.5 What is an ion trap?

## SUBELEMENT 4BG - Practical Circuits (6 Questions)

4BG-1.1 What is a flip-flop circuit?
4BG-1.2 How many bits of information can be stored in a single flip-flop circuit?
4BG-1.3 What is a bistable multivibrator circuit?
4BG-1.4 What is an astable multivibrator?
4BG-1.5 What is a monostable multivibrator?
4BG-1.6 What is the schematic symbol for an AND gate?
4BG-1.7 What is the schematic symbol for an OR gate?
4BG-1.8 What is the schematic symbol for a NAND gate?
4BG-1.9 What is the schematic symbol for a NOR gate?
4BG-1.10 What is the schematic symbol for a NOT gate?
4BG-2.1 How is elecronic division usually accomplished?
4BG-2.2 How many output changes are obtained for every two impulses applied to the input of a bistable flip-flop circuit?
4BG-2.3 What is a crystal marker generator?
4BG-2.4 What additional circuitry is required in a 100-kHz crystal marker generator to provide markers at 50- and 25-kHz?
4BG-2.5 What is the purpose of a prescaler circuit?
4BG-2.6 What does the accuracy of a counter depend upon?
4BG-2.7 Why is 1 MHz the standard crystal reference in many counter circuits?
4BG-2.8 How many flip-flops are required to divide a signal by 4?
4BG-3.1 What are the advantages of using an op-amp over LC elements in an audio filter?
4BG-3.2 What determines the gain and frequency characteristics of an op-amp LRC active filter?
4BG-3.3 What are the principle uses of an op-amp RC active filter in amateur circuitry?
4BG-3.4 What type of capacitors should be used in an op-amp RC active filter circuit?
4BG-3.5 How can unwanted ringing and audio instability be prevented in a multisection op-amp LRC audio filter circuit?
4BG-3.6 Where should an op-amp RC active filter be placed in an amateur receiver?

4BG-3.7 What is meant by the term <u>offset</u> voltage with regard to an op-amp?

4BG-3.8 What is the term for the temperature coefficient of offset voltage with respect to time an an op-amp?

4BG-3.9 What is meant by the term <u>popcorn</u> noise in an op-amp?

4BG-3.10 What is meant by the term <u>flicker</u> noise in an op-amp?

4BG-4.1 What is meant by the term <u>noise figure</u> of a communications receiver?

4BG-4.2 What is the term used to express the degree to which a communications receiver approaches the quietness of a theoretical receiver?

4BG-4.3 Which stage of a receiver primarily establishes its noise figure?

4BG-4.4 What is the limiting condition for sensitivity in a communications receiver?

4BG-4.5 What is the theoretical minimum noise floor of a receiver with a 400 Hz bandwidth?

4BG-5.1 How can selectivity be employed in the front end circuitry of a communications receiver?

4BG-5.2 What degree of selectivity is desirable in the i-f circuitry of an amateur ssb receiver?

4BG-5.3 What degree of selectivity is desirable in the i-f circuitry of an amateur telegraphy receiver?

4BG-5.4 What degree of selectivity is desirable in the i-f circuitry of an amateur rtty receiver?

4BG-5.5 What degree of selectivity is desirable in the i-f circuity of an amateur FM receiver.

4BG-5.6 What degree of selectivity is desirable in the audio circuitry of an amateur ssb receiver?

4BG-5.7 What degree of selectivity is desirable in the audio circuitry of an amateur telegraphy receiver?

4BG-5.8 What degree of selectivity is desirable in the audio circuitry of an amateur rtty receiver?

4BG-5.9 What degree of selectivity is desirable in the audio circuitry of an amateur voice FM receiver?

4BG-5.10 How can selectivity be employed in the local oscillator circuitry of a communications receiver?

4BG-6.1 What is meant by the <u>dynamic range</u> of a communications signal?

4BG-6.2 What is the term for the ratio between the largest tolerable receiver input signal and the minimum discernable signal?

4BG-6.3 What is the optimum dynamic range obtainable with the present state of the receiver art?

4BG-6.4 What type of problems are cause by poor dynamic range in a communications receiver?

4BG-6.5 The ability of a communications receiver to perform well in the presence of strong signals outside the amateur band of interest is indicated by what figure?

4BG-7.1 What voltage gain can be expected from the circuit shown in Figure 4BG-7.1, when R1 is 1000-ohms and Rf is 100-kohms?

4BG-7.2 What voltage gain can be expected from the circuit shown in Figure 4BG-7.1, when R1 is 1800-ohms and Rf is 68-kohms?

4BG-7.3 What voltage gain can be expected from the circuit shown in Figure 4BG-7.1, when R1 is 3300-ohms and Rf is 47-kohms?

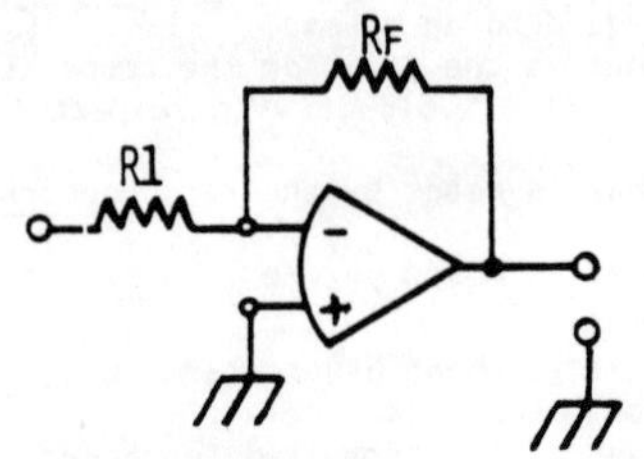

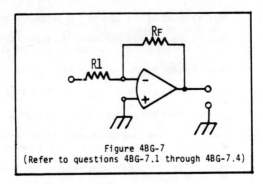

Figure 4BG-7
(Refer to questions 4BG-7.1 through 4BG-7.4)

4BG-7.4 What voltage gain can be expected from the circuit shown in Figure 4BG-7.1, when Rl is 10-kohms and Rf is 47-kohms?

4BG-7.5 What parameters must be selected when designing a audio filter using an op-amp?

4BG-8.1 In an amplifier employing an FET, what determines the input impedance?

4BG-8.2 In an amplifier employing an FET, what determines the output impedance?

4BG-9.1 What frequency range will be tuned by the preselector circuit shown in Figure 4BG-9 when L is 10-microhenrys, Cf is 156-picofarads, and Cv is 50-picofarads maximum and 2-picofarads minimum?

4BG-9.2 What frequency range will be tuned by the preselector circuit shown in Figure 4BG-9 when L is 30 microhenrys, Cf is 200-picofarads, and Cv is 80-picofarads maximum and 10-picofarads mimimum?

4BG-9.3 What is the purpose of a bypass capacitor in a circuit?

4BG-9.4 What is the purpose of a coupling capacitor in a circuit?

4BG-9.5 How does the selection of bypass and coupling capacitors in a single stage amplifier affect its frequency response?

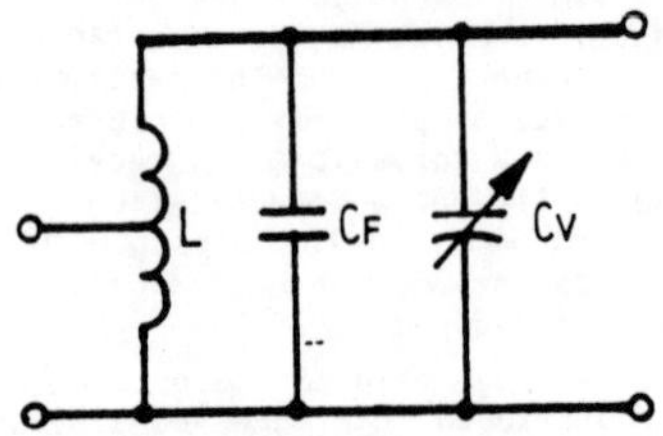

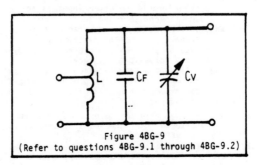

Figure 4BG-9
(Refer to questions 4BG-9.1 through 4BG-9.2)

SUBELEMENT 4BH - Signals and Emissions (3 Questions)

4BH-1.1 In a pulse-width modulation system, what does

the modulating signal vary?

4BH-1.2 What is the type of modulation where the modulating signal varies the duration of the transmitted pulse?

4BH-1.3 In a pulse-position modulation system, what does the modulating signal vary?

4BH-1.4 Why is the transmitter peak power in a pulse modulation system much greater than its average power?

4BH-1.5 How is voice transmitted in a pulse modulation system?

4BH-2.1 What is the duration of a 60 speed Baudot RTTY data pulse?

4BH-2.2 What is the duration of a 60 speed Baudot RTTY start pulse?

4BH-2.3 What is the duration of a 60 speed Baudot stop pulse?

4BH-2.4 What is the meaning of the term baud rate?

4BH-2.5 What is the baud rate for a 60 speed Baudot RTTY system?

4BH-3.1 What is amplitude compandored single sideband?

4BH-3.2 What is meant by compandoring?

4BH-3.3 What is the purpose of a pilot tone in an amplitude compandored single sideband system?

4BH-3.4 What is the approximate frequency of the pilot tone in an amplitude comandored single sideband system?

4BH-3.5 How many more voice transmissions can be packed into a give frequency band for amplitude compandored single sideband systems over conventional FM?

4BH-4.1 What is the bandwidth of a 170 Hz shift, 60 speed Baudot RTTY, F1 transmission?

4BH-4.2 What is the bandwidth of a 170 Hz shift, 100 speed Baudot RTTY, F1 transmission?

4BH-4.3 What is the bandwidth of a 20 wpm international Morse code A1 transmission?

4BH-4.4 What is the bandwidth of a 170 Hz shift, 1200 baud rate ASCII, F1 transmission?

4BH-4.5 What is the bandwidth of a 170 Hz shift, 19,200 baud rate ASCII, F1 transmission?

4BH-4.6 What is the bandwidth of a 5 wpm international Morse code A1 transmission?

4BH-4.7 What is the bandwidth of a 170 Hz shift, 110 baud rate ASCII, F1 transmission?

4BH-4.8 What is the bandwidth of a 170 Hz shift, 75 speed Baudot RTTY, F1 transmission?

4BH-4.9 What is bandwidth of a 13 wpm international Morse code A1 transmission?

4BH-4.10 What is the bandwidth of a 170 Hz shift, 300 baud rate ASCII, F1 transmission?

4BH-5.1 What is the term for the amplitude of the maximum positive excursion of a signal?

4BH-5.2 What is the term for the amplitude of the maximum negative excursion of a signal?

4BH-6.1 When viewing a pure sine wave signal on a oscilloscope, what is the easiest voltage amplitude dimension to measure?

4BH-6.2 In a symetrical wave form, what is the relationship between the pk-pk voltage and the peak voltage amplitude?

4BH-6.3 What amplitude parameter is valuable in evaluating the signal-handling capability of linear processing device?

SUBELEMENT 4BI - Antennas and Feedlines (5 Questions)

4BI-1.1 What factors determine the receiving antenna gain at an amateur radio station in earth operation?
4BI-1.2 What factors determine the EIRP required by an amateur radio station in earth operation?
4BI-1.3 What factors determine the EIRP required by an amateur radio station in telecommand operation?
4BI-1.4 What is the change in the gain of a parabolic dish type antenna when the operating frequency is doubled?
4BI-1.5 What happens to the bandwidth of an antenna as the gain is increased?
4BI-1.6 How can circular polarization be produced using linearly polarized antennas?
4BI-1.7 What is the beamwidth of an antenna which produces a symmetrical pattern and has a gain of 20?
4BI-1.8 What is the beamwidth of an antenna which produces a symmetrical pattern and has a gain of 30?
4BI-1.9 What is the beamwidth of an antenna which produces a symmetrical pattern and has a gain of 15?
4BI-1.10 What is the beamwidth of an antenna which produces a symmetrical pattern and has a gain of 12?
4BI-2.1 What is an isotropic radiator?
4BI-2.2 What would be the antenna pattern for an isotropic radiator?
4BI-2.3 What is the directivity of an isotropic radiator?
4BI-2.4 When is it useful to refer to an isotropic radiator?
4BI-2.5 What are the two most commonly used standards for comparing the properties of actual antennas used by amateurs?
4BI-2.6 What can be said about the directivity pattern of an isotropic antenna?
4BI-2.7 How much gain does a half-wave dipole have over an isotropic radiator?
4BI-2.8 What purpose does an isotropic radiator serve?
4BI-2.9 A certain antenna has a 6 dB gain over an isotropic antenna. How much gain does the antenna have over a half-wave dipole?
4BI-2.10 A certain antenna has 12 dB gain over an isotropic antenna. How much gain does the antenna have area a half-wave dipole?
4BI-3.1 What kind of radiation pattern results when two 1/4 wavelength vertical antennas spaced 1/2 wavelength apart are fed 180 degrees out of phase?
4BI-3.2 What kind of radiation pattern results when two 1/4 wavelength vertical antennas spaced 1/4 wavelength apart are fed 90 degrees out of phase?
4BI-3.3 What kind of radiation pattern results when two 1/4 wavelength vertical antennas spaced 1/2 wavelength apart are fed in phase?
4BI-3.4 How far apart should two 1/4 wavelength vertical antennas be spaced in order to produce a figure-eight pattern that is broadside to the plane of the verticals, when the antennas are fed in phase?
4BI-3.5 How far should two 1/4 wavelength vertical antennas be spaced in order to produce a figure-

eight pattern that is in line with the vertical antennas, when the antennas are fed 180 degrees out of phase?

4BI-3.6 What kind of radiation pattern results when two 1/4 wavelength vertical antennas spaced 1/4 wavelength apart are fed 180 degrees out of phase?

4BI-3.7 What kind of radiation pattern results when two 1/4 wavelength vertical antennas spaced 1/8 wavelength apart are fed 180 degrees out of phase?

4BI-3.8 What kind of radiation pattern results when two 1/4 wavelength vertical antennas spaced 1/8 wavelength apart are fed in phase?

4BI-3.9 What kind of radiation pattern results when two 1/4 wavelength vertical antennas spaced 1/4 wavelength apart are fed in phase?

4BI-3.10 What kind of radiation pattern results when two 1/4 wavelength vertical antennas spaced 1/4 wavelength apart are fed 90 degrees out of phase?

4BI-4.1 What is a nonresonant rhombic antenna?

4BI-4.2 What is a resonant rhombic antenna?

4BI-4.3 What is the effect of using a terminating resistor on a rhombic antenna?

4BI-4.4 What should be the value of the terminating resistor on a rhombic antenna?

4BI-4.5 What are the advantages of a nonresonant rhombic antenna?

4BI-4.6 What are the disadvantages of a nonresonant rhombic antenna?

4BI-4.7 What is the characteristic impedance of an ordinary rhombic antenna, looking into the input end?

4BI-5.1 What is the delta matching system for matching an antenna to a feedline?

4BI-5.2 What is the gamma matching system for matching an antenna to a feedline?

4BI-5.3 What is the stub system for matching an antenna to a feedline?

4BI-5.4 What should be the maximum capacitance of the resonating capacitor in a gamma matching circuit for a 20 meter dipole?

4BI-5.5 What should be the maximum capacitance of the resonating capacitor in a gamma matching circuit for a 10 meter dipole?

4BI-6.1 What does a 1/4 wavelength of transmission line look like to a generator when the line is shorted at the far end?

4BI-6.2 What does a 1/4 wavelength of transmission line look like to a generator when the line is open at the far end?

4BI-6.3 What does a 1/8 wavelength of transmission line look like to a generator when the line is shorted at the far end?

4BI-6.4 What does a 1/8 wavelength of transmission line look like to a generator when the line is open at the far end?

4BI-6.5 What does a 1/2 wavelength of transmission line look like to a generator when the line is shorted at the far end?

4BI-6.6 What does a 1/2 wavelength of transmission line look like to a generator when the line is open at the far end?

4BI-6.7 What does a 3/8 wavelength of transmission line

look like to a generator when the line is shorted at the far end?

4BI-6.8 What does a 3/8 wavelength of transmission line look like to a generator when the line is open at the far end?

# Appendix B

# Part 97

**Subpart A — General**

Sec.

97.1 Basis and purpose.
97.3 Definitions.

**Subpart B — Amateur Operator and Station Licenses**

*Operator Licenses*

97.5 Classes of operator licenses.
97.7 Privileges of operator licenses.
97.9 Eligibility for new operator license.
97.11 Application for operator license.
97.13 Renewal or modification of operator license.

*Operator License Examinations*

97.19 When examination is required.
97.21 Examination elements.
97.23 Examination requirements.
97.25 Examination credit.
97.26 Examination procedure.
97.27 Examination preparation
97.28 Examination administration.
97.29 Examination grading.
97.31 Volunteer Examiner requirements.
97.33 Volunteer Examiner conduct.

**Subpart C — Technical Standards**

97.61 Authorized frequencies and emissions.
97.63 Selection and use of frequencies.
97.65 Emission limitations.
97.67 Maximum authorized power.
97.69 Digital communication.
97.73 Purity of emissions.
97.75 Use of external radio frequency (RF) power amplifiers.
97.76 Requirements for type acceptance of external radio frequency (RF) power amplifiers and external radio frequency power amplifier kits.
97.77 Standards for type-acceptance of external radio frequency (RF) power amplifiers and external radio frequency power amplifier kits.

**Subpart D — Operating Requirements and Procedures**

*General*

97.78 Practice to be observed by all licensees.
97.79 Control operator requirements.

97.35 Temporary operating authority.

*Station Licenses*

97.37 General eligibility for station license.
97.39 Eligibility of corporations or organizations to hold station license.
97.40 Station license required.
97.41 Operation of Canadian Amateur Radio stations in the U.S..
97.42 Application for station license.
97.43 Mailing address furnished by licensee.
97.44 Location of station.
97.45 Limitations on antenna structures.
97.47 Renewal and/or modification of amateur station license.
97.49 Commission modification of station license.

*Call Signs*

97.51 Assignment of call signs.

*Duplicate Licenses and License Term*

97.57 Duplicate license.
97.59 License term.
97.81 Authorized apparatus.
97.82 Availability of operator license.
97.83 Availability of station license.
97.84 Station identification.
97.85 Repeater operation.
97.86 Auxiliary operation.
97.87 Beacon operation.
97.88 Operation of a remotely controlled station.
97.89 Points of communications.
97.90 System Network Diagram required.
97.91 One-way communications.
97.92 Record of operations.
97.93 Modulation of carrier.

*Station Operation Away from authorized Location*

97.95 Operation away from the authoized fixed station location.

*Special Provisions*

97.99 Stations used only for radio control of remote model crafts and vehicles.
97.101 Mobile stations aboard ships or aircraft.

*Emergency Operations*

97.107 Operation in emergencies.

**Subpart E — Prohibited Practices and Administrative Sanctions**

*Prohibited Transmission and Practices*

97.110 Business communications prohibited.
97.111 Limitations on international communications.
97.112 No remuneration for use of station.
97.113 Broadcasting prohibited.
97.114 Third-party traffic.
97.115 Music prohibited.
97.116 Amateur radiocommunication for unlawful purposes prohibited.
97.117 Codes and ciphers prohibited.
97.119 Obscenity, indecency, profanity.
97.121 False signals.
97.123 Unidentified communications.
97.125 Interference.
97.126 Retransmitting radio signals.
97.127 Damage to apparatus.
97.129 Fraudulent licenses.

*Administrative Sanctions*

97.131 Restricted operation.
97.133 Second notice of same violation.
97.135 Third notice of same violation.
97.137 Answers to notices of violation.

**Subpart F — Radio Amateur Civil Emergency Service (RACES)**

97.307 Issuance of permit.
97.309 Modification, suspension, or cancellation of permit.
97.311 Operating conditions.
97.313 Station identification.

**Subpart H — Amateur-Satellite Service**

*General*

97.401 Purposes.
97.403 Definitions.
97.405 Applicability of rules.
97.407 Eligibility for space operation.
97.409 Eligibility for earth operation.
97.411 Eligibility for telecommand operation.
97.413 Space operations requirements.

*Technical Requirements*

97.415 Frequencies available.

*Special Provisions*

97.417 Space operation.
97.419 Telemetry.
97.421 Telecommand operation.
97.422 Earth operation.
97.423 Notification required.

**Subpart I — Volunteer-Examiner Coordinators**

*General*

97.501 Purpose.
97.503 Definitions.

*General*

97.161 Basis and purpose.
97.163 Definitions.
97.165 Applicability of rules.

*Station Authorizations*

97.169 Station license required.
97.171 Eligibility for RACES station license.
97.173 Application for RACES station license.
97.175 Amateur radio station registration in civil defense organization.

*Operating Requirements*

97.177 Operator requirements.
97.179 Operator privileges.
97.181 Availability of RACES station license and operator licenses.

*Technical Requirements*

97.185 Frequencies available.
97.189 Points of communications.
97.191 Permissible communications.
97.193 Limitations on the use of RACES stations.

**Subpart G — Operation of Amateur Radio Stations in the United States by Aliens Pursuant to Reciprocal Agreements**

97.301 Basis, purpose, and scope.
97.303 Permit required.
97.305 Application for permit.
97.505 Applicability of rules.
97.507 VEC qualifications.

*Volunteer-Examiner Coordinator Functions*

97.511 Agreement required.
97.513 Scheduling of examination.
97.515 Coordinating volunteer examiners.
97.517 Examinations.
97.519 Examination procedures.
97.521 Evaluation of questions.

*Appendices*

1 Examination points.
2 Extracts from Radio Regulations Annexed to the International Telecommunication Convention (Geneva, 1959).
3 Classification of emissions.
4 Convention between the United States of America and Canada, Relating to the Operation by Citizens of Either Country of Certain Radio Equipment or Stations in the Other Country (Effective May 15, 1952).
5 Determination of Antenna Height above Average Terrain.

Authority: §§97.1 to 97.313 issued under 48 Stat. 1066, 1082, as amended; 47 U.S.C. 154, 303. Interpret or apply 48 Stat. 1064-1068, 1081-1105, as amended; 47 U.S.C. Sub-chap. I, III-VI.

# Subpart A — General

**§ 97.1 Basis and purpose.**

The rules and regulations in this part are designed to provide an amateur radio service having a fundamental purpose as expressed in the following principles:

(a) Recognition and enhancement of the value of the amateur service to the public as a voluntary noncommercial communication service, particularly with respect to providing emergency communications.

(b) Continuation and extension of the amateur's proven ability to contribute to the advancement of the radio art.

(c) Encouragement and improvement of the amateur radio service through rules which provide for advancing skills in both the communication and technical phases of the art.

(d) Expansion of the existing reservoir within the amateur radio service of trained operators, technicians, and electronics experts.

(e) Continuation and extension of the amateur's unique ability to enhance international goodwill.

amateur radio communication.

(f) *Primary station.* The principal amateur radio station at a specific land location shown on the station license.

(g) *Military recreation station.* An amateur radio station licensed to the person in charge of a station at a land location provided for the recreational use of amateur radio operators, under military auspices of the Armed Forces of the United States.

(h) *Club station.* A separate Amateur radio station licensed to an amateur radio operator acting as a station trustee for a *bona fide* amateur radio organization or society. A *bona fide* Amateur Radio organization or society shall be composed of at least two persons, one of whom must be a licensed amateur operator, and shall have:

(1) A name,

(2) An instrument of organization (e.g., constitution),

(3) Management, and

(4) A primary purpose which is devoted to Amateur Radio activities consistent with §97.1 and constituting the major portion of the club's activities.

**§ 97.3 Definitions.**

(a) *Amateur radio service.* A radio communication service of self-training, intercommunication, and technical investigation carried on by amateur radio operators.

(b) *Amateur radio communication.* Noncommercial radio communication by or among amateur radio stations solely with a personal aim and without pecuniary or business interest.

(c) *Amateur radio operator* means a person holding a valid license to operate an amateur radio station issued by the Federal Communications Commission.

(d) *Amateur radio license.* The instrument of authorization issued by the Federal Communications Commission comprised of a station license, and in the case of the primary station, also incorporating an operator license.

*Operator license.* The instrument of authorization including the class of operator privileges.

*Interim Amateur Permit.* A temporary operator and station authorization issued to licensees successfully completing Commission-supervised examinations for higher class operator licenses.

*Station license.* The instrument of authorization for a radio station in the Amateur Radio Service.

(e) *Amateur radio station.* A station licensed in the amateur radio service embracing necessary apparatus at a particular location used for

(i) (Reserved)

(j) *Terrestrial location.* Any point within the major portion of the earth's atmosphere, including aeronautical, land and maritime locations.

(k) (Reserved)

(l) *Amateur radio operation.* Amateur radio communication conducted by amateur radio operators from amateur radio stations, including the following:

*Fixed operation.* Radio communication conducted from the specific geographical land location shown on the station license.

*Portable operation.* Radio communication conducted from a specific geographical location other than that shown on the station license.

*Mobile operation.* Radio communication conducted while in motion or during halts at unspecified locations.

*Repeater operation.* Radio communication, other than auxiliary operation, for retransmitting automatically the radio signals of other amateur radio stations.

*Auxiliary operation.* Radio communication for remotely controlling other amateur radio stations, for automatically relaying the radio signals of other amateur radio stations in a system of stations, or for intercommunicating with other amateur radio stations in a system of amateur radio stations.

*Beacon operation.* One-way radio communication conducted in order to facilitate measurement of radio equipment characteristics, adjustment of radio equipment, observation of propagation or transmission phenomena, or other related experimental activities.

*Radio control operation*. One-way radio communication for remotely controlling objects or apparatus other than amateur radio stations.

(m) *Control* means techniques used for accomplishing the immediate operation of an Amateur Radio station. Control includes one or more of the following:

(1) *Local control*. Manual control, with the control operator monitoring the operation on duty at the control point located at a station transmitter with the associated operating adjustments directly accessible. (Direct mechanical control, or direct wire control of a transmitter from a control point located on board any aircraft, vessel, or on the same premises on which the transmitter is located, is also considered local control.)

(2) *Remote control*. Manual control, with the control operator monitoring the operation on duty at a control point located elsewhere than at the station transmitter, such that the associated operating adjustment are accessible through a control link.

(RF) power generated by operations of an amateur radio station, including the following:

(1) *Transmitter power*. The peak envelope power (output) present at the antenna terminals (where the antenna feedline, or if no feedline is used, the antenna, would be connected) of the transmitter. The term "transmitter" includes any external radio frequency power amplifier which may be used. Peak envelope power is defined as the average power during one radio frequency cycle at the crest of the modulation envelope, taken under normal operating conditions.

(2) *Effective radiated power*. The product of the transmitter (peak envelope) power, expressed in watts, delivered to an antenna, and the relative gain of the antenna over that of a half-wave dipole antenna.

(u) *System network diagram*. A diagram showing each station and its relationship to the other stations in a network of stations, and to the control point(s).

(v) *Third-party traffic*. Amateur radio communication by or under the supervision of the control operator at an amateur radio station to another amateur radio station on behalf of anyone other than the control operator.

(w) *Emergency communication*. Any amateur radio communication directly relating to the im-

(3) *Automatic control* means the use of devices and procedures for control so that a control operator does not have to be present at the control point at all times.

(n) *Control link*. Apparatus for effecting remote control between a control point and a remotely controlled station.

(o) *Control operator*. An amateur radio operator designated by the licensee of an amateur radio station to also be responsible for the emissions from that station.

(p) *Control point*. The operating position of an amateur radio station where the control operator function is performed.

(q) *Antenna structures*. Antenna structures include the radiating system, its supporting structures, and any appurtenances mounted thereon.

(r) *Antenna height above average terrain*. The height of the center of radiation of an antenna above an averaged value of the elevation above sea level for the surrounding terrain.

(s) *Transmitter*. Apparatus for converting electrical energy received from a source into radio-frequency electromagnetic energy capable of being radiated.

(t) *Transmitting power*. The radio frequency

mediate safety of life of individuals or the immediate protection of property.

(x) *Automatic retransmission*. Retransmission of signals by an amateur radio station whereby the retransmitting station is actuated solely by the presence of a received signal through electrical or electro-mechanical means, i.e., without any direct, positive action by the control operator.

(y) *External radio frequency power amplifier*. Any device which, (1) when used in conjunction with a radio transmitter as a signal source, is capable of amplification of that signal, and (2) is not an integral part of the transmitter as manufactured.

(z) *External radio frequency power amplifier kit*. Any number of electronic parts, usually provided with a schematic diagram or printed circuit board, which, when assembled in accordance with instructions, results in an external radio frequency power amplifier, even if additional parts of any type are required to complete assembly.

(aa) (Reserved)

(bb) *Business communications*. Any transmission or communication the purpose of which is to facilitate the regular business or commercial affairs of any party.

# Subpart B — Amateur Operator and Station Licenses

**§ 97.5 Classes of operator licenses**

Amateur Extra class.
Advanced class (previously class A).
General class (previously class B).
Conditional class (previously class C).
Technician class.
Novice class.

**§ 97.7 Privileges of operator licenses.**

(a) *Amateur Extra Class and Advanced Class.* All authorized amateur privileges including exclusive frequency operating authority in accordance with the following table:

| *Frequencies* | *Class of license authorized* |
|---|---|
| 3500-3525 kHz | Amateur Extra Only |
| 3750-3775 kHz | |
| 7000-7025 kHz | |
| 14,000-14,025 kHz | |
| 14,150-14,175 kHz | |
| 21,000-21,025 kHz | |
| 21,200-21,225 kHz | |

**§ 97.11 Application for operator license.**

(a) An application (FCC Form 610) for a new operator license, including an application for change in operating privileges, which will require an examination shall be submitted in accordance with the provisions of §97.26.

(b) An application (FCC Form 610) for renewal and/or modification of license when no change in operating privileges is involved shall be submitted to the Commission's office at Gettysburg, Pennsylvania 17325.

**§ 97.13 Renewal or modification of operator license.**

(a) An amateur operator license may be renewed upon proper application.

(b) The applicant shall qualify for a new license by examination if the requirements of this section are not fulfilled.

(c) Application for renewal and/or modification of an amateur operator license shall be submitted on FCC Form 610 and shall be accompanied by the applicant's license or a photocopy

| | |
|---|---|
| 3775-3850 kHz<br>7150-7225 kHz<br>14,175-14,225 kHz<br>21,225-21,300 kHz | Amateur Extra and Advanced. |

(b) *General Class.* All authorized amateur privileges except those exclusive operating privileges which are reserved to the Advanced Class and/or Amateur Extra Class.

(c) *Conditional Class.* Same privileges as General Class. New Conditional Class licenses will not be issued. Present Conditional Class licensees will be issued General Class licenses at time of renewal or modification.

(d) *Technician Class.* All authorized amateur privileges on the frequencies 50.0 MHz and above. Technician Class licenses also convey the full privileges of Novice Class licenses.

(e) *Novice Class.* Radiotelegraphy in the frequency bands 3700-3750 kHz, 7100-7150 kHz (7050-7075 kHz when terrestrial station location is not within Region 2), 21,100-21,200 kHz, and 28,100-28,200 kHz, using only type A1 emission and using only the international Morse code.

**§ 97.9 Eligibility for new operator license.**

Anyone except a representative of a foreign government is eligible for an amateur operator license.

thereof. Application for renewal of unexpired licenses must be made during the license term and should be filed within 90 days, but not later than 30 days, prior to the end of the license term. In any case in which the licensee has, in accordance with the provisions of this chapter, made timely and sufficient application for renewal of an unexpired license, no license with reference to any activity of a continuing nature shall expire until such application shall have been finally determined.

(d) If a license is allowed to expire, application for renewal may be made during a grace period of two years after the expiration date. During this grace period, an expired license is not valid. A license renewed during the grace period will be dated currently and will not be backdated to the date of its expiration. Application for renewal shall be submitted on FCC Form 610 and shall be accompanied by the applicant's expired license or a photocopy thereof.

## OPERATOR LICENSE EXAMINATIONS

**§ 97.19 When examination is required.**

Examination is required for the issuance of a new amateur operator license, and for a change in class of operating privileges. Credit may be given, however, for certain elements of examination as provided in §97.25.

**§ 97.21 Examination elements.**

Examinations for amateur operator privileges will comprise one or more of the following examination elements:

(a) Element 1(A): Beginner's code test at five (5) words per minute;

(b) Element 1(B): General code test at thirteen (13) words per minute;

(c) Element 1(C): Expert's code test at twenty (20) words per minute;

(d) Element 2: Basic law comprising rules and regulations essential to beginners operation, including sufficient elementary radio theory for the understanding of those rules;

(e) Element 3: General amateur practice and regulations involving radio operation and apparatus and provisions of treaties, statutes, and rules affecting amateur stations and operators.

(f) Element 4(A): Intermediate amateur practice involving intermediate level radio theory and operation as applicable to modern amateur techniques, including, but not limited to, radiotelephony and radiotelegraphy;

(g) Element 4(B): Advanced amateur practice involving advanced radio theory and operation as applicable to modern amateur techniques, including, but not limited to, radiotelephony, radiotelegraphy, and transmissions of energy for measurements and observations applied to propexamination credit, this certificate is valid for a period of one year from the date of its issuance.

(c) A person who applies for an amateur operator license will be given credit for any telegraphy element if that person holds a commercial radiotelegraph operator license or permit issued by the Federal Communications Commission, or has held one within 5 years of the Commission's receipt of that person's application for an amateur operator license.

(d) No examination credit, except as herein provided, shall be allowed on the basis of holding or having held any amateur or commercial operator license.

**§ 97.26 Examination procedure.**

(a) Each examination for an amateur radio operator license shall be administered at a location and a time specified by the examiner(s). Public announcement before examinations shall be made for elements 1(B), 1(C), 3, 4(A) and 4(B).

(b) The examiner(s) must be present and observing the candidate throughout the entire examination.

(c) The examiner(s) will be responsible for the proper conduct and necessary supervision during each examination.

(d) Each candidate for an amateur radio

agation, for the radio control of remote objects and for similar experimental purposes.

**§ 97.23 Examination requirements.**

Applicants for operator licenses will be required to pass the following examination elements:

(a) Amateur Extra Class: Elements 1(C), 2, 3, 4(A) and 4(B);

(b) Advanced Class: Elements 1(B), 2, 3, 4(A);

(c) General Class: Elements 1(B), 2 and 3;

(d) Technician Class: Elements 1(A), 2 and 3;

(e) Novice Class: Elements 1(A) and 2.

**§ 97.25 Examination credit.**

(a) An applicant for a higher class of amateur operator license who holds any valid amateur license will be required to pass only those elements of the higher class examination that are not included in the examination for the amateur license held.

(b) A certificate of successful completion of an examination will be issued to applicants who successfully complete an examination element. Upon presentation of this certificate for telegraphy examination elements 1(A), 1(B) or 1(C), examiners shall give the applicant for an amateur radio operator license examination credit for the code speed associated with the previously completed element. For purposes of license, which requires the applicant to pass one or more examination elements, must present the examiner(s) with a properly completed FCC Form 610 on or before the registration deadline date for those examination sessions for which registration is required; otherwise, applicants shall submit FCC Form 610 at the examination session before the start of the examination(s). In cases where a registration deadline is required, it shall be specified by the VEC that issues the examination papers to the examiner.

(e) The candidate shall comply with the instructions given by the examiner(s). The examiner(s) must immediately terminate the examination upon failure of the candidate to comply with the examiner(s)' instructions.

(f) At the completion of the examination, the candidate shall return all test papers to the examiner(s).

(g) A candidate whose physical disabilities require special procedures to allow participation in examination sessions shall attach a statement to his/her application. For examinations other than Novice Class the statement shall be retained in the files of the VEC that issues the test papers. The statement for Novice Class examinations shall be retained by the examiner for one year. The statement shall include:

(1) a physician's certification indicating the nature of the disability; *AND*

(2) the name(s) of the person(s) taking and transcribing the applicant's dictation of test questions and answers, if such a procedure is necessary.

(h) An applicant who fails an examination element required for an amateur radio operator license shall not apply to be examined for the same or higher examination element within thirty days of the date the examination element was failed.

### § 97.27 Examination preparation.

(a) Element 1(A) shall be prepared by the examiner. The preparer must hold an Amateur Extra, Advanced, or General Class operator license. The test shall be such as to prove the applicant's ability to transmit correctly by hand key, straight key, or, if supplied by the applicant, any other type of hand operated key such as a semi-automatic or electronic key, but not a keyboard keyer, and to receive correctly by ear texts in the international Morse code at a rate of not less than five (5) words per minute. (Special procedures may be employed in cases of physical disability. See §97.26(g).) The applicant is responsible for knowing, and may be tested on, the twenty-six letters of the alphabet, the numerals 0-9, the period, the comma, the question mark, $\overline{AR}$, $\overline{SK}$, $\overline{BT}$ and $\overline{DN}$. (See §97.29(c).)

(b) Elements 1(B) and 1(C) shall be prepared the appropriate PR Bulletin in the form in which they have been approved by the Commission. Beginning January 1, 1987, volunteer examiners may also design Elements 3, 4(A), and 4(B) in accord with the provisions of this paragraph. Each VEC and each volunteer examiner is required to hold current examination designs in confidence.

(e) PR Bulletins 1035A, B, and C and D will be composed of questions originated by the FCC and questions submitted by amateur radio operators in accordance with the instructions in the Bulletin. Amateur radio operators holding Amateur Extra Class licenses may submit questions for any written examination element. Amateur radio operators holding Advanced Class licenses may only submit questions for Element 2 and 3. Amateur radio operators holding General Class or Technician Class licenses may only submit questions for Element 2.

### § 97.28 Examination administration.

(a) Unless otherwise prescribed by the Commission, each examination for an amateur radio operator license (except the Novice Class operator license) shall be administered by three accredited (see §97.515) volunteer examiners. An examiner administering telegraphy examination Element 1(A) or written examination Element 2 (in conjunction with an examination other than

by the examiners or be obtained by the examiners from the VEC. The preparer must hold an Amateur Extra Class license. The test shall be such as to prove the applicant's ability to transmit correctly by hand key, straight key, or, if supplied by the applicant, any other type of hand operated key such as a semi-automatic or electronic key, but not a keyboard keyer, and to receive correctly by ear texts in the international Morse code at not less than the prescribed speed. (Special procedures may be employed in cases of physical disability. See §97.26(g).) The applicant is responsible for knowing, and may be tested on, the twenty-six letters of the alphabet, the numerals 0-9, the period, the comma, the question mark, $\overline{AR}$, $\overline{SK}$, $\overline{BT}$ and $\overline{DN}$. See §97.29(c).

(c) Element 2 shall be designed by the examiner from PR Bulletin 1035A (latest date of issue), entitled *Questions for the Element 2 Amateur Radio Operator License Examination.*

(d) Elements 3, 4(A) and 4(B) will be designed by the VEC. The VEC will select questions for each test from the appropriate list of questions approved by the Commission (either PR Bulletin 1035B, C, or D, latest date of issue). The VEC must select the appropriate number of questions from each category of the syllabus (PR Bulletin 1035) as specified in PR Bulletin 1035 B, C, or D. These questions must be taken verbatim from a Novice Class examination) or written examination Element 3 must hold an Amateur Extra Class or Advanced Class radio operator license. An examiner administering telegraphy examination Element 1(B) or 1(C) or written examination Element 4(A) or 4(B) must hold an Amateur Extra Class radio operator license.

(b) Unless otherwise prescribed by the Commission, each examination for the Novice Class operator license shall be administered by one volunteer examiner. The examiner does not have to be accredited. The volunteer examiner must hold a current General, Advanced or Amateur Extra Class operator license issued by the Commission.

(c) Upon completion of an examination element, the examiner(s) shall immediately grade the test papers.

(d) When the candidate does not score a passing grade on an examination element, the examiner(s) shall so inform the candidate by providing the percentage of questions answered correctly, and by returning the application (see §97.26) to the candidate. For examinations other than Novice Class examinations, the test papers, including answer sheets, shall be returned to the VEC that issued them. For Novice Class examinations, the test papers, including answer sheets, must be retained as part of the volunteer examiner's station records for one year from the date the examination is administered.

(e) When the candidate scores a passing grade on an examination element, the examiner(s) (except for examinations for the Novice Class operator license) must issue a certificate of successful completion of the examination. This certificate may be used for a period of one year for examination credit for telegraphy elements 1(A), 1(B) or 1(C) (see 97.25[b].

(f) When the candidate scores a passing grade on all examination elements required for the operator license class sought (see §97.23), the examiners shall certify to the following information on the candidate's application form (see §97.26):

(1) Examiners' names, addresses and amateur radio station call signs;

(2) Examiners' qualifications to administer the examination; (see §97.31); *AND*

(3) Examiners' signed statements that the applicant has passed the required examination elements.

(g) Within ten days of the administration of a successful examination for the Novice Class operator license, the examiner shall submit the candidate's application to:

Federal Communications Commission
Gettysburg, Pennsylvania 17325

(h) Within ten days of the administration of a successful examination for the Technician,

(c) An applicant passes a code element examination if he/she proves his/her ability to transmit correctly by hand key (straight key, or, if supplied by the applicant, any other type of hand operated key such as a semi-automatic or electronic key, but not a keyboard keyer) and to receive correctly by ear texts in the international Morse code at not less than the prescribed speed during a five-minute test period. Each five characters shall be counted as one word. Each punctuation mark and numeral shall be counted as two characters.

**§ 97.31 Volunteer examiner requirements.**

(a) Each volunteer examiner administering an examination for an amateur radio operator license must:

(i) be at least 18 years of age; *AND*

(ii) not be related to the candidate.

b) Any person who owns a significant interest in, or is an employee of, any company or other entity which is engaged in the manufacture or distribution of equipment used in connection with amateur radio transmissions, or in the preparation or distribution of any publication used in preparation for obtaining amateur station operator licenses, is ineligible to be a volunteer examiner for purposes of administering an amateur radio operator examination. However, a person who does not normally com-

General, Advanced or Amateur Extra Class operator license, the examiners shall submit the successful candidates' applications and all test papers to the VEC that originally issued that test.

(i) The FCC reserves the right, without qualification, to:

(1) administer examinations itself; *OR*

(2) readminister examinations itself or under the supervision of an examiner designated by the FCC, to any person who obtained an operator license above the Novice Class through the volunteer examination process.

(j) If a licensee fails to appear for readministration of an examination pursuant to paragraph (i)(2) of this section, or does not successfully complete the examination element(s) which are readministered, the licensee's operator license is subject to cancellation; in an instance of such cancellation, the licensee will be issued an operator license consistent with the completed examination elements which have not been invalidated by not appearing for or failing readministration of an examination.

### § 97.29 Examination grading.

(a) Each examination element shall be graded separately by the examiners.

(b) An applicant passes a written examination if he/she answers at least 74 percent of the questions correctly.

municate with that part of an entity engaged in the manufacture or distribution of such equipment, or in the preparation or distribution of any publication used in preparation for obtaining amateur operator licenses, is eligible to be a volunteer examiner.

(c) Volunteer examiners may not be compensated for services. They may be reimbursed for out-of-pocket expenses, except for Novice class examinations (see §97.36).

(d) Each volunteer examiner administering an examination for the Technician, General, Advanced or Amateur Extra Class operator license must be accredited by the Volunteer-Examiner Coordinator (see Subpart I).

(e) The FCC will not accept the services of any person seeking to be a volunteer examiner if that person's amateur radio station license or amateur radio station operator's license has ever been revoked or suspended.

### § 97.33 Volunteer examiner conduct.

No volunteer examiner shall give or certify any examination by fraudulent means or for monetary or other consideration. Violation of this provision may result in the revocation of the amateur radio station license and the suspension of the amateur radio operator license of the volunteer examiner. This does not preclude a volunteer examiner from accepting reimbursement for out-of-pocket expenses under §97.36. Reimbursement in any amount in excess of that permitted may result in the sanctions specified herein.

### § 97.35 Temporary operating authority.

Unless the FCC otherwise prescribes, an applicant already licensed in the Amateur Radio Service, upon successfully completing the amateur radio examination(s) required for a higher class, may operate an amateur radio station consistent with the rights and privileges of the higher class for a period of one year from the date of the most recently completed examination for that operator class provided that the applicant retains the certificate(s) for successful completion of the examination(s) (see §97.28[e]) at the staton location, provided that the applicant uses the identifier code of the new class of license for which the applicant has qualified (KT for Technician Class, AG for General Class, AA for Advanced Class and AE for Amateur Extra Class) as a suffix to the present call sign (see §97.84), and provided that the FCC has not yet acted upon the application for a higher class of license.

### §97.36 Reimbursement for expenses.

(a) Each volunteer examiner coordinator and each volunteer examiner may be reimbursed by

(d) The expense and reimbursement records must be retained by each volunteer examiner coordinator and each volunteer examiner for 3 years and made available to the FCC upon request.

(e) Each volunteer examiner must forward on or before January 15 of each year the certification concerning expenses to the volunteer examiner coordinator who coordinated the efforts of the volunteer examiner and for which reimbursement was received. The volunteer examiner coordinator must forward all such certifications and its own certification concerning expenses to the FCC on or before January 31 of each year.

(f) The volunteer examiner coordinator must disaccredit any volunteer examiner who fails to provide the annual certification. The volunteer examiner coordinator must advise the FCC on January 31 of each year of the volunteer examiners that it has disaccredited for this reason.

## STATION LICENSES

### § 97.37 General eligibility for station license.

(a) An Amateur Radio station license will be issued only to a licensed Amateur Radio operator, except that a military recreation station license may also be issued to an individual not licensed as an Amateur Radio operator

examinees for out-of-pocket expenses incurred in preparing, processing or administering examinations for amateur station operator licenses above the Novice class. The volunteer examiner coordinator or the volunteer examiners must collect the reimbursement fee, if any, from the examinees. No reimbursement may be accepted for preparing, processing or administering Novice class examinations.

(b) The maximum amount of reimbursement is $4.00 for 1984 and will be adjusted annually each January 1 thereafter for changes in the Department of Labor Consumer Price Index. Changes in the maximum amount of reimbursement will be announced by the Commission in a Public Notice. The amount of such reimbursement fee from any examinee for any one examination at a particular session regardless of the number of examination elements taken must not exceed the published maximum.

(c) Each volunteer examiner coordinator and each volunteer examiner who accepts reimbursement must maintain records of out-of-pocket expenses and reimbursements for each examination session. They must certify on or before January 31 of each year to the Commission's office in Gettysburg, PA 17325 that all expenses for the period from January 1 to December 31 of the preceding year for which reimbursement was obtained were necessarily and prudently incurred.

(other than a representative of a foreign government), who is in charge of a proposed military recreation station not operated by the U.S. Government but which is to be located in approved public quarters.

(b) Only modification and/or renewal station licenses will be issued for club and military recreation stations. No new licenses will be issued for these types of stations.

**§ 97.39 Eligibility of corporations or organizations to hold station license.**

An amateur station will not be issued to a school, company, corporation, association, or other organization, except that in the case of a *bona fide* Amateur Radio organization or society meeting the criteria set forth in Section 97.3, a station license may be issued to a licensed amateur operator, other than the holder of a Novice class license, as trustee for such society.

**§ 97.40 Station license required.**

(a) No transmitting station shall be operated in the Amateur Radio Service without being licensed by the Federal Communications Commission, except that an Amateur Radio station licensed by the Government of Canada may, in accordance with Section 97.41, be operated in the United States without the prior approval of the Commission.

(b) Every Amateur Radio operator shall have one, but only one, primary Amateur Radio station license.

**§ 97.41 Operation of Canadian amateur stations in the United States.**

(a) An Amateur Radio station licensed by the Government of Canada may be operated in the United States without the prior approval of the Federal Communications Commission.

(b) Operation of a Canadian amateur station in the United States must comply with all of the following:

(1) The terms of the Convention between the United States and Canada (TIAS No. 2508) relating to the operation by citizens of either country of certain radio equipment or stations in the other country. (See Appendix 4 to Part 97.)

(2) The operating terms and conditions of the amateur station license issued by the Government of Canada.

(3) The provisions of subpart A through E of Part 97.

(4) Any further conditions the Commission may impose upon the privilege of operating in the United States.

(c) At any time the Commission may, in its discretion, modify, suspend, or cancel the privilege of any Canadian licensee operating an Amateur Radio station in the United States.

license to operate in the vicinity of an FCC monitoring station are advised to give consideration, prior to filing applications, to the possible need to protect the FCC stations from harmful interference. Geographical coordinates of the facilities which require protection are listed in Section 0.121(c) of the Commission's Rules. Applications for stations (except mobile stations) in the vicinity of monitoring stations may be reviewed by Commission staff on a case-by-case basis to determine the potential for harmful interference to the monitoring station. Depending on the theoretical field strength value and existing root-sum-square or other ambient radio field signal levels at the indicated coordinates, a clause protecting the monitoring station may be added to the station license.

(2) Advance consultation with the Commission is suggested prior to filing an initial application for station license if the proposed station will be located within one mile of any of the above-referenced monitoring station coordinates and is to be operated on frequencies below 1000 MHz. Such consultations are also suggested for proposed stations operating above 1000 MHz if they are to be located within one mile of any monitoring station designated in Section 0.121(c) as a satellite monitoring facility.

(3) Regardless of any coordination prior to filing initial applications, it is suggested that

**§ 97.42 Application for station license.**

(a) Each application for a club or military recreation station license in the Amateur Radio Service shall be made on the FCC Form 610-B. Each application for any other American Radio license shall be made on the FCC Form 610.

(b) One application and all papers incorporated therein and made a part thereof shall be submitted for each amateur station license. If the application is only for a station license, it shall be filed directly with the Commission's Gettysburg, Pennsylvania office. If the application also contains an application for any class of amateur operator license, it shall be filed in accordance with the provisions of §97.11.

(c) Each applicant in the Private Radio Services (1) for modification of a station license involving a site change or a substantial increase in tower height or (2) for a license for a new station must, before commencing construction, supply the environmental information, where required, and must follow the procedure prescribed by Subpart 1 of Part 1 of this chapter (§§ 1.1301 through 1.1319) unless Commission action authorizing such construction would be a minor action within the meaning of Subpart 1 of Part 1.

(d) Protection for Federal Communications Commission Monitoring Stations:

(1) Applicants for an Amateur Radio station licensees within one mile of a monitoring station consult the Commission before initiating any changes in the station which would increase the field strength produced over the monitoring station.

(4) Applicants and licensees desiring such consultations should communicate with: Chief, Field Operation Bureau, Federal Communications Commission, Washington, DC 20054, Telephone 202-632-6980.

(5) The Commission will not screen applications to determine whether advance consultation has taken place. However, applicants are advised that such consultation can avoid objections from the Federal Communications Commission or modification of any authorization which will cause harmful interference.

**§ 97.43 Mailing address furnished by licensee.**

Each application shall set forth and each licensee shall furnish the Commission with an address in the United States to be used by the Commission in serving documents or directing correspondence to that licensee. Unless any licensee advises the Commission to the contrary, the address contained in the licensee's most recent application will be used by the Commission for this purpose.

**§ 97.44 Location of station.**

Every amateur radio station shall have one land location, the address of which appears in the station license, and at least one control point.

**§ 97.45 Limitations on antenna structures.**

(a) Except as provided in paragraph (b) of this section, an antenna for a station in the Amateur Radio Service which exceeds the following height limitations may not be erected or used unless notice has been filed with both the FAA on FAA Form 7460-1 and with the Commission on Form 854 or on the license application form, and prior approval by the Commission has been obtained for:

(1) Any construction or alteration of more than 200 feet in height above ground level at its site (§ 17.7[a] of this chapter).

(2) Any construction or alteration of greater height than an imaginary surface extending outward and upward at one of the following slopes (§17.7[b] of this chapter):

(i) 100 to 1 for a horizontal distance of 20,000 feet from the nearest point of the nearest runway of each airport with at least one runway more than 3,200 feet in length, excluding heliports and seaplane bases without specified boundaries, if that airport is either listed in the Airport Directory of the current Airman's Information Manual or is operated by a Federal military agency.

Applicants claiming such exemption shall submit a statement with their application to the Commission explaining the basis in detail for their finding (§17.14[a] of this chapter).

(2) Any antenna structure of 20 feet or less in height except one that would increase the height of another antenna structure (§17.14[b] of this chapter).

(c) Further details as to whether an aeronautical study and/or obstruction marking and lighting may be required, and specifications for obstruction marking and lighting when required, may be obtained from Part 17 of this chapter, "Construction, Marking and Lighting of Antenna Structures." Information regarding the inspection and maintenance of antenna structures requiring obstruction marking and lighting is also contained in Part 17 of this chapter.

**§ 97.47 Renewal and/or modification of amateur station license.**

(a) Application for renewal and/or modification of an individual station license shall be submitted on FCC Form 610, and application for renewal and/or modification of an amateur club or military recreation station shall be submitted on FCC Form 610-B. In every case the application shall be accompanied by the applicant's license or photocopy thereof. Applications for renewal of unexpired licenses must be made during the license term and should be filed not later

(ii) 50 to 1 for a horizontal distance of 10,000 feet from the nearest point of the nearest runway of each airport with its longest runway no more than 3,200 feet in length, excluding heliports and seaplane bases without specified boundaries, if that airport is either listed in the Airport Directory or is operated by a Federal military agency.

(iii) 25 to 1 for a horizontal distance of 5,000 feet from the nearest point of the nearest landing and takeoff area of each heliport listed in the Airport Directory or operated by a Federal military agency.

(3) Any construction or alteration on an airport listed in the Airport Directory of the Airman's Information Manual (§17.7[c] of this chapter).

(b) A notification to the Federal Aviation Administration is not required for any of the following construction or alteration:

(1) Any object that would be shielded by existing structures of a permanent and substantial character or by natural terrain or topographic features of equal or greater height, and would be located in the congested area of a city, town, or settlement where it is evident beyond all reasonable doubt that the structure so shielded will not adversely affect safety in air navigation.

than 60 days prior to the end of the license term. In any case in which the licensee has in accordance with the provisions of this chapter, made timely and sufficient application for renewal of an unexpired license, no license with reference to any activity of a continuing nature shall expire until such application shall have been finally determined.

(b) If a license is allowed to expire, application for renewal may be made during a grace period of two years after the expiration date. During this grace period, an expired license is not valid. A license renewal during the grace period will be dated currently and will not be backdated to the date of its expiration. An application for an individual station license shall be submitted on FCC Form 610. An application for an amateur club or military recreation station license shall be submitted on FCC Form 610-B. In every case the application shall be accompanied by the applicant's expired license or a photocopy thereof.

**§ 97.49 Commission modification of station license.**

(a) Whenever the Commission shall determine that the public interest, convenience, and necessity would be served, or any treaty ratified by the United States will be more fully complied with, by the modification of any radio station license either for a limited time, or for the duration of the term thereof, it shall issue an order for such licensee to show cause why such license should not be modified.

(b) Such order to show cause shall contain a statement of the grounds and reasons for such proposed modification, and shall specify wherein the said license is required to be modified. It shall require the licensee against whom it is directed to appear at a place and time therein named, in no event to be less than 30 days from the date of receipt of the order, to show cause why the proposed modification should not be made and the order of modification issued.

(c) If the licensee against whom the order to show cause is directed does not appear at the time and place provided in said order, a final order of modification shall issue forthwith.

## CALL SIGNS

### § 97.51 Assignment of call signs.

(a) The Commission shall assign the call sign of an amateur radio station on a systematic basis.

(b) The Commission shall not grant any request for a specific call sign.

(c) From time to time the Commission will issue public announcements detailing the policies and procedures governing the systematic assignment of call signs and any changes in those policies and procedures.

## DUPLICATE LICENSES AND LICENSE TERM

### § 97.57 Duplicate license.

Any licensee requesting a duplicate license to replace an original which has been lost, mutilated, or destroyed, shall submit a statement setting forth the facts regarding the manner in which original license was lost, mutilated, or destroyed. If, subsequent to receipt by the licensee of the duplicate license, the original license is found, either the duplicate or the original license shall be returned immediately to the Commission.

### § 97.59 License term.

(a) Amateur operator licenses are normally valid for a period of ten years from the date of issuance of a new, modified or renewed license.

(b) Amateur station licenses are normally valid for a period of ten years from the date of issuance of a new, modified or renewed license. All amateur station licenses, regardless of when issued, will expire on the same date as the licensee's amateur operator license.

(c) A duplicate license shall bear the same expiration date as the license for which it is a duplicate.

# Subpart C — Technical Standards

**§ 97.61 Authorized frequencies and emissions.**

(a) The following frequency bands and associated emissions are available to amateur radio stations for amateur radio operation, other than repeater operation, auxiliary operation and automatically-controlled beacon operation, subject to the limitations of §97.65 and paragraph (b) of this section:

| Frequency band *(kHz)* | Emissions | Limitations (See paragraph[b]) |
|---|---|---|
| 1800-2000 | A1,A3 | |
| 3500-4000 | A1 | |
| 3500-3750 | F1 | |
| 3750-4000 | A3,A4,A5,F3,F4,F5 | 4 |
| 5167.5 | A3A,A3J | 13 |
| 7000-7300 | A1 | 3,4 |
| 7000-7150 | F1 | 3,4 |
| 7075-7100 | A3,F3 | 11 |
| 7150-7300 | A3,A4,A5,F3,F4,F5 | 3,4 |
| 14000-14350 | A1 | |
| 14000-14150 | F1 | |
| 14150-14350 | A3,A4,A5,F3,F4,F5 | |
| 21000-21450 | A1 | |
| 21000-21200 | F1 | |
| 21200-21450 | A3,A4,A5,F3,F4,F5 | |
| 28000-29700 | A1 | |
| 28000-28300 | F1 | |
| 28300-29700 | A3,A4,A5,F3,F4,F5 | |

| Frequency band (MHz) | Emissions | Limitations (See paragraph[b]) |
|---|---|---|
| 50.0-54.0 | A1 | |
| 50.1-54.0 | A2,A3,A4,A5,F1,F2,F3,F4,F5 | |
| 51.0-54.0 | A0,F0 | |
| 144.0-148.0 | A1 | |
| 144.1-148.0 | A0,A2,A3,A4,A5,F0,F1,F2,F3,F4,F5 | |
| 220-225 | A0,A1,A2,A3,A4,A5,F0,F1,F2,F3,F4,F5 | 5 |
| 420-450 | A0,A1,A2,A3,A4,A5,F0,F1,F2,F3,F4,F5 | 5,7 |
| 1215-1300 | A0,A1,A2,A3,A4,A5,F0,F1,F2,F3,F4,F5 | 5 |
| 2300-2310 | N0N,A1A,A2A,A3B,A3E,A3C,A3F,F1B,F2B,F3E, G3E,F3C,F3F,P0N | 5,8 |
| 2390-2450 | N0N,A1A,A2A,A3B,A3E,A3C,A3F,F1B,F2B,F3E,G3E, F3C,F3F,P0N | 5,8 |
| (GHz) | | |
| 3.300-3.500 | A0,A1,A2,A3,A4,A5,F0,F1,F2,F3,F4,F5,P | 5,12 |
| 5.650-5.925 | A0,A1,A2,A3,A4,A5,F0,F1,F2,F3,F4,F5,P | 5,9 |
| 10.000-10.500 | A0,A1,A2,A3,A4,A5,F0,F1,F2,F3,F4,F5 | 5 |
| 24.000-24.250 | A0,A1,A2,A3,A4,A5,F0,F1,F2,F3,F4,F5,P | 5,10 |
| 48.000-50.000 | A0,A1,A2,A3,A4,A5,F0,F1,F2,F3,F4,F5,P | |
| 71.000-76.000 | A0,A1,A2,A3,A4,A5,F0,F1,F2,F3,F4,F5,P | |
| 165.000-170.000 | A0,A1,A2,A3,A4,A5,F0,F1,F2,F3,F4,F5,P | |
| 240.000-250.000 | A0,A1,A2,A3,A4,A5,F0,F1,F2,F3,F4,F5,P | |
| Above 300.000 | A0,A1,A2,A3,A4,A5,F0,F1,F2,F3,F4,F5,P | |

(b) Limitations:
(1) (Reserved)
(2) (Reserved)

(3) Where, in adjacent regions or subregions, a band of frequencies is allocated to different

(iv) Those portions of California and Nevada south of latitude 37° 10′ N., and the area within

services of the same category, the basic principle is the equality of right to operate. Accordingly, the stations of each service in one region or subregion must operate so as not to cause harmful interference to services in the other regions or subregions (No. 117, the Radio Regulations, Geneva, 1959).

(4) 3900-4000 kHz and 7100-7300 kHz are not available in the following U.S. possessions: Baker, Canton, Enderbury, Guam, Howland, Jarvis, the Northern Mariana Islands, Palmyra, American Samoa, and Wake Islands.

(5) Amateur stations shall not cause interference to the Government radiolocation service.

(6) (Reserved)

(7) In the following areas the peak envelope power output of a transmitter used in the Amateur Radio Service shall not exceed 50 watts, except when authorized by the appropriate Commission Engineer-in-Charge and the appropriate Military Area Frequency Coordinator.

(i) Those portions of Texas and New Mexico bounded by latitude 33° 24′ N., 31° 53′., and longitude 105° 40′ W. and 106° 40′ W.

(ii) The State of Florida, including the Key West area and the areas enclosed within circles of 200-mile radius centered at 28° 21′ N., 80° 43′ W. and 30° 30′ N., 86° 30′ W.

(iii) The State of Arizona.

a 200-mile radius of 34° 09′ N., 119° 11′ W.

(v) In the State of Massachusetts within a 160-kilometer (100 miles) radius of 41° 45′ N., 70° 32′ W.

(vi) In the State of California within a 240-kilometer (150 miles) radius of 39° 08′ N., 121° 26′ W.

(vii) In the State of Alaska within a 160-kilometer (100 miles) radius of 64° 17′ N., 149° 10′ W.

(viii) In the State of North Dakota within a 160-kilometer (100 miles) radius of 48° 43′ N., 97° 54′ W.

(ix) In the States of Alabama, Florida, Georgia and South Carolina within a 200 kilometer (124 mile) radius of Warner Robins Air Force Base, Georgia (latitude 32° 38′ North, longitude 83° 35′ West).

(x) In the State of Texas within a 200 kilometer (124 mile) radius of Goodfellow Air Force Base, Texas (latitude 31° 25′ North, longitude 100° 24′ West).

(8) No protection in the band 2400-2500 MHz is afforded from interference due to the operation of industrial, scientific, and medical devices on 2450 MHz.

(9) No protection in the band 5725-5875 MHz is afforded from interference due to the operation of industrial, scientific and medical devices on 5800 MHz.

**Authorized Effective Radiated Power for Repeater Stations**

| *Antenna height above average terrain in meters* | *Maximum effective radiated power for frequency bands above:* | | |
|---|---|---|---|
| | *29.5 MHz* | *420 MHz* | *1215 MHz* |
| Below 32 (105 feet) | 800 watts | Paragraphs (a) and (b) | Paragraphs (a) and (b) |
| 32-160 (105-525 feet) | 400 watts | 800 watts | ..do.. |
| 160-320 (525-1050 feet) | 200 watts | 800 watts | ..do.. |
| Above 320 (1050 feet) | 100 watts | 400 watts | ..do.. |

(10) No protection in the band 24.00-24.25 GHz is afforded from interference due to the operation of industrial, scientific and medical devices on 24.125 GHz.

(11) The use of A3 and F3 in this band is limited to amateur radio stations located outside Region 2 and amateur radio stations located within Region 2 which are west of 130 degrees West longitude.

(12) Amateur stations shall not cause interference to the Fixed-Satellite Service operating in the band 3400-3500 MHz.

(13) The frequency 5167.5 kHz, maximum power 150 watts, may be used by any station authorized under this part to communicate with any other station authorized in the State of under paragraph (a) of this section. Limitations of paragraph (b) of this section apply.

**§ 97.63 Selection and use of frequencies.**

(a) An amateur station may transmit on any frequency within any authorized amateur frequency band.

(b) Sideband frequencies resulting from keying or modulating a carrier wave shall be confined within the authorized amateur band.

(c) The frequencies available for use by a control operator of an amateur station are dependent on the operator license classification of the control operator and are listed in §97.7.

**§ 97.65 Emission limitations.**

Alaska for emergency communications. All stations operating on this frequency must be located in or within 50 nautical miles of the State of Alaska. The frequency 5167.5 kHz may be used by licensees in the Alaska-private fixed service for calling and listening, but only for establishing communication before switching to another frequency.

(c) All amateur frequency bands above 29.5 MHz are available for repeater operation, except 50.0-52.0 MHz, 144.0-144.5 MHz, 145.5-146.0 MHz, 220.0-220.5 MHz, 431.0-433.0 MHz, and 435.0-438.0 MHz. Both the input (receiving) and output (transmitting) frequencies of a station in repeater operation shall be frequencies available for repeater operation.

(d) All amateur frequency bands above 220.5 MHz, except 431-433 MHz, and 435-438 MHz, are available for auxiliary operation.

(e) The following amateur frequency bands and emissions are available for automatically-controlled beacon operation: 28.20-28.30 MHz, 50.06-50.08 MHz, 144.05-144.06 MHz, 220.05-220.06 MHz, 222.05-222.06 MHz and 432.07-432.08 MHz using type AØ, A1, FØ, F1 or A2J emissions (when type F1 or A2J emissions are employed in these bands, the radio or audio frequency shift, as appropriate, shall not exceed 1000 Hz); all amateur frequency bands above 450 MHz using emission types authorized

(a) Type AØ emission, where not specifically designated in the bands listed in §97.61, may be used for short periods of time when required for authorized remote control purposes or for experimental purposes. However, these limitations do not apply where type AØ emission is specifically designated.

(b) Whenever code practice, in accordance with §97.91(d), is conducted in bands authorized for A3 emission tone modulation of the radiotelephone transmitter may be utilized when interspersed with appropriate voice instructions.

(c) On frequencies below 29.0 MHz, the bandwidth of an F3 emission (frequency or phase modulation) shall not exceed that of an A3 emission having the same audio characteristics.

(d) On frequencies below 50 MHz, the bandwidth of A4, A5, F4 and F5 emissions shall not exceed that of an A3 single-sideband emission.

(e) On frequencies between 50 MHz and 225 MHz:

(1) The bandwidth of A4 and A5 single-sideband emissions shall not exceed the bandwidth of an A3 single-sideband emission.

(2) The bandwidth of A4 and A5 double-sideband emissions shall not exceed the bandwidth of an A3 double-sideband emissions.

(3) F4 and F5 emissions shall utilize a peak carrier deviation no greater than 5 kHz and a maximum modulating frequency no greater than 3

kHz or, alternatively, shall occupy a bandwidth no greater than 20 kHz. (For this purpose the bandwidth is defined as the width of the frequency band, outside of which the mean power of any emission is attenuated by at least 26 decibels below the mean power level of the total emission. A 3 kHz sampling bandwidth is used by the FCC in making this determination.)

(f) Below 225 MHz, an A3 emission may be used simultaneously with an A4 or A5 emission on the same carrier frequency, provided that the total bandwidth does not exceed that of an A3 double-sideband emission.

### § 97.67 Maximum authorized transmitting power.

(a) Notwithstanding other limitations of this section, amateur radio stations shall use the minimum transmitting power necessary to carry out the desired communications.

(b) Each amateur radio transmitter may be operated with a peak envelope power output (transmitter power) not exceeding 1500 watts, except as provided in paragraph (a) of this section. Other limitations of this section and §97.61 also apply.

(c) Within the limitations of paragraphs (a) and (b) of this section, the effective radiated power of an amateur radio station in repeater operation shall not exceed the power specified

radio communication may include digital codes which represent alphanumeric characters, analogue measurements or other information. These digital codes may be used for such communications as (but not limited to) radio teleprinter and other objects, transference of computer programs or direct computer-to-computer communications, and communications in various types of data networks (including so-called "packet switching" systems); provided that such digital codes are not intended to obscure the meaning of, but are only to facilitate, the communications, and further provided that such operation is carried out in accordance with other regulations set forth in this part. (For purposes of this section, the sending speed (signaling rate), in baud, is defined as the reciprocal of the shortest (signaling) time interval (in seconds) that occurs during a transmission, where each time interval is the period between changes of transmitter state (including changes in emission amplitude, frequency, phase, or combination of these, as authorized).)

(a) The use of the digital codes specified in paragraph (b) of this section is permitted on any amateur frequency where F1 emission is permitted, subject to the following requirements:

(1) The sending speed shall not exceed the following:

(i) 300 baud on frequencies below 28 MHz;

for the antenna height above average terrain [in the table above.]

(d) The peak envelope power output (transmitter power) of each amateur radio transmitter shall not exceed 200 watts when transmitting in any of the following frequency bands:

(1) 3700-3750 kHz;

(2) 7100-7150 kHz (7050-7075 kHz when the terrestrial location of the station is within Region 1 or 3);

(3) 21100-21200 kHz;

(4) 28100-28200 kHz.

(e) Within the limitations of paragraph (a) of this section, the peak envelope power output of an amateur radio station in beacon operation shall not exceed 100 watts.

(f) An amateur radio station may transmit A3 emissions on or before June 1, 1990 with a transmitter power exceeding that authorized by paragraph (b) of this section, provided that the power input (both radio frequency and direct current) to the final amplifying stage supplying radio frequency power to the antenna feedline does not exceed 1000 watts, exclusive of power for heating the cathodes of vacuum tubes. Limitations of paragraphs (a), (c) and (d) of this section and limitations of §97.61 still apply.

**§ 97.69 Digital communications.**

Subject to the special conditions contained in paragraphs (a), (b) and (c) below, an amateur

(ii) 1200 baud on frequencies between 28 and 50 MHz;

(iii) 19.6 kilobaud on frequencies between 50 and 220 MHz;

(iv) 56 kilobaud on frequencies above 220 MHz.

(2) When type A2, F1 or F2 emissions are used on frequencies below 50 MHz, the radio or audio frequency shift (the difference between the frequency for the "mark" signal and that for the "space" signal), as appropriate, shall not exceed 1000 Hz. When these emissions are used on frequencies above 50 MHz, the frequency shift, in hertz, shall not exceed the sending speed, in baud, of the transmission, or 1000 Hz, whichever is greater.

(b) Except as provided for in paragraph (c) of this section, only the following digital codes, as specified, may be used:

(1) The International Telegraph Alphabet Number 2 (commonly known as Baudot); provided that transmission shall consist of a single channel, five unit (start-stop) teleprinter code conforming to the International Telegraph Alphabet Number 2 with respect to all letters and numerals (including the slant sign or fraction bar); however, in the "figures" positions not utilized for numerals, special signals may be employed for the remote control of receiving printers, or for other purposes indicated in this section.

(2) The American Standard Code for Information Interchange (commonly known as ASCII); provided that the code shall conform to the American Standard Code for Information Interchange as defined in American National Standards Institute (ANSI) Standard X3.4-1968.

(3) The International Radio Consultative Committee (CCIR) Recommendations 476-2 and 476-3 (commonly known as AMTOR); provided that the code, baud rate and emission timing shall conform to the specifications of CCIR 476-2 (1978) or CCIR 476-3 (1982), Mode A or Mode B.

(c) In addition to the above provisions, the use of any digital code is permitted on amateur frequencies above 50 MHz, except those on which only A1 emission is permitted, subject to the following requirements:

(1) Communications using such digital codes are authorized for domestic operation only (communications between points within areas where radio services are regulated by the U.S. Federal Communications Commission), except when special arrangements have been made between the United States and the administration of any other country concerned.

(2) The bandwidth of an emission from a station using such digital codes shall not exceed the following (where for this purpose the bandwidth is defined as the width of the frequency band,

**§ 97.71 (Reserved).**

**§ 97.73 Purity of emissions.**

(a) Except for a transmitter or transceiver built before April 15, 1977 or first marketed before January 1, 1978, the mean power of any spurious emission or radiation from an amateur transmitter, transceiver, or external radio frequency power amplifier being operated with a carrier frequency below 30 MHz shall be at least 40 decibels below the mean power of the fundamental without exceeding the power of 50 milliwatts. For equipment of mean power less than five watts, the attenuation shall be at least 30 decibels.

(b) Except for a transmitter or transceiver built before April 15, 1977 or first marketed before January 1, 1978, the mean power of any spurious emission or radiation from an amateur transmitter, transceiver, or external radio frequency power amplifier being operated with a carrier frequency above 30 MHz but below 225 MHz shall be at least 60 decibels below the mean power of the fundamental. For a transmitter having a mean power of 25 watts or less, the mean power of any spurious radiation supplied to the antenna transmission line shall be at least 40 decibels below the mean power of the fundamental without exceeding the power of 25 microwatts, but need not be reduced below the power of 10 microwatts.

(c) Paragraphs (a) and (b) of this section not-outside of which the mean power of any emission is attenuated by at least 26 decibels below the mean power of the total emission; a 3 kHz sampling bandwidth being used by the FCC in making this determination):

(i) 20 kHz on frequencies between 50 and 220 MHz;

(ii) 100 kHz on frequencies between 220 and 1215 MHz;

(iii) On frequencies above 1215 MHz any bandwidth may be used provided that the emission is in accordance with §97.63(b) and §97.73(c).

(3) A description of the digital code and the modulation technique shall be included in the station log during all periods of use and shall be provided to the Commission on request.

(4) When deemed necessary by an Engineer-in-charge of a Commission field facility to assure compliance with the rules of this part, a station licensee shall:

(i) Cease the transmission of digital codes authorized under this paragraph.

(ii) Restrict the transmission of digital codes authorized under this paragraph to the extent instructed.

(iii) Maintain a record, convertible to the original information (voice, text, image, etc.), of all coded communications transmitted under authority of this paragraph.

withstanding, all spurious emissions or radiation from an amateur transmitter, transceiver, or external radio frequency power amplifier shall be reduced or eliminated in accordance with good engineering practice.

(d) If any spurious radiation, including chassis or power line radiation, causes harmful interference to the reception of another radio station, the licensee may be required to take steps to eliminate the interference in accordance with good engineering practice.

NOTE: For the purpose of this section, a spurious emission or radiation means any emission or radiation from a transmitter, transceiver, or external radio frequency power amplifier which is outside of the authorized Amateur Radio Service frequency band being used.

**§ 97.74 (Reserved).**

**§ 97.75 Use of external radio frequency (RF) power amplifiers.**

(a) Any external radio frequency (RF) power amplifier used or attached at any amateur radio station shall be type accepted in accordance with Subpart J of Part 2 of the FCC's Rules for operation in the Amateur Radio Service, unless one or more of the following conditions are met:

(1) The amplifier is not capable of operation on any frequency or frequencies below 144 MHz

(the amplifier shall be considered incapable of operation below 144 MHz if the mean output power decreases, as frequency decreases from 144 MHz, to a point where 0 decibels or less gain is exhibited at 120 MHz and below and the amplifier is not capable of being easily modified to provide amplification below 120 MHz);

(2) The amplifier was originally purchased before April 28, 1978;

(3) The amplifier was —

(i) Constructed by the licensee, not from an external RF power amplifier kit, for use at his amateur radio station;

(ii) Purchased by the licensee as an external RF power amplifier kit before April 28, 1978 for use at his amateur radio station; or

(iii) Modified by the licensee for use at his amateur radio station in accordance with §2.1001 of the FCC's Rules;

(4) The amplifier was purchased by the licensee from another amateur radio operator who —

(i) Constructed the amplifier, but not from an external RF power amplifier kit;

(ii) Purchased the amplifier as an external RF power amplifier kit before April 28, 1978 for use at his amateur radio station; or

(iii) Modified the amplifier for use at his amateur radio station in accordance with §2.1001 of the FCC's Rules;

(5) The external RF power amplifier was purchased (as defined in §2.815), manufactured, imported or modified for use in the Amateur Radio Service shall be type accepted for use in the Amateur Radio Service in accordance with Subpart J of Part 2 of the FCC's Rules. This requirement does not apply if one or more of the following conditions are met:

(1) The amplifier is not capable of operation on any frequency or frequencies below 144 MHz. For the purpose of this part, an amplifier will be deemed to be incapable of operation below 144 MHz if the amplifier is not capable of being easily modified to increase its amplification characteristics below 120 MHz, and either;

(i) The mean output power of the amplifier decreases, as frequency decreases from 144 MHz, to a point where 0 decibels or less gain is exhibited at 120 MHz and below 120 MHz; or

(ii) The amplifier is not capable of even short periods of operation below 120 MHz without sustaining permanent damage to its amplification circuitry.

(2) The amplifier was originally purchased before April 28, 1978 by an amateur radio operator for use at his amateur radio station;

(3) The amplifier was constructed or modified by an amateur radio operator for use at his amateur radio station in accordance with §2.1001 of the FCC's Rules;

(4) The amplifier was constructed or modified

chased from a dealer who obtained it from an amateur radio operator who —

(i) Constructed the amplifier, but not from an external RF power amplifier kit;

(ii) Purchased the amplifier as an external RF power amplifier kit before April 28, 1978, for use at his amateur radio station; or

(iii) Modified the amplifier for use at his amateur radio station in accordance with §2.1001 of the FCC's Rules; or

(6) The amplifier was originally purchased after April 28, 1978 and has been issued a marketing waiver by the FCC.

(b) A list of type-accepted equipment may be inspected at FCC headquarters in Washington, DC or at any FCC field office. Any external RF power amplifier appearing on this list as type accepted for use in the Amateur Radio Service may be used in the Amateur Radio service.

NOTE: No more than one unit of one model of an external RF power amplifier shall be constructed or modified during any calendar year by an amateur radio operator for use in the Amateur Radio Service without a grant of type acceptance.

**§ 97.76 Requirements for type acceptance of external radio frequency (RF) power amplifiers and external radio frequency power amplifier kits.**

(a) Any external radio frequency (RF) power amplifier or external RF power amplifier kit by an amateur radio operator in accordance with §2.1001 of the FCC's Rules and sold to another amateur radio operator or to a dealer;

(5) The amplifier is purchased in used condition by an equipment dealer from a licensed amateur radio operator who constructed or modified the equipment in accordance with §2.1001 of the FCC's Rules and the amplifier is further sold to another amateur radio operator for use at his/her licensed amateur radio station.

(6) The amplifier was manufactured before April 28, 1978 and has been issued a marketing waiver by the FCC.

(b) No more than one unit of one model of an external RF power amplifier shall be constructed or modified during any calendar year by an amateur radio operator for use in the Amateur Radio Service without a grant of type acceptance.

(c) A list of type-accepted equipment may be inspected at FCC headquarters in Washington, DC or at any FCC field office. Any external RF power amplifier appearing on this list as typeaccepted for use in the Amateur Radio Service may be marketed for use in the Amateur Radio Service.

**§ 97.77 Standards for type acceptance of external radio frequency (RF) power amplifiers and external radio frequency power amplifier kits.**

(a) An external radio frequency (RF) power

amplifier or external RF power amplifier kit will receive a grant of type acceptance under this Part only if a grant of type acceptance would serve the public interest, convenience or necessity.

(b) To receive a grant of type acceptance under this Part, an external RF power amplifier shall meet the emission limitations of §97.73 when the amplifier is —

(1) Operated at its full output power;

(2) Placed in the "standby" or "off" positions, but still connected to the transmitter; and

(3) Driven with at least 50 watts mean radio frequency input power (unless a higher drive level is specified).

(c) To receive a grant of type acceptance under this part, an external RF power amplifier shall not be capable of operation on any frequency or frequencies between 24.00 MHz and 35.00 MHz. The amplifier will be deemed incapable of operation between 24.00 MHz and 35.00 MHz if —

(1) The amplifier has no more than 6 decibels of gain between 24.00 MHz and 26.00 MHz and between 28.00 and 35.00 MHz. (This gain is determined by the ratio of the input RF driving signal (mean power measurement) to the mean RF output power of the amplifier); and

(2) The amplifier exhibits no amplification (0 decibels of gain) between 26.00 MHz and 28.00 MHz.

the amplifier's operating characteristics in a manner contrary to the FCC's Rules;

(3) Instructions for operation or modification of the amplifier in a manner contrary to the FCC's Rules;

(4) Any internal or external controls or adjustments to facilitate operation of the amplifier in a manner contrary to the FCC's Rules;

(5) Any internal radio frequency sensing circuitry or any external switch, the purpose of which is to place the amplifier in the transmit mode;

(6) The incorporation of more gain in the amplifier than is necessary to operate in the Amateur Radio Service. For purposes of this paragraph, an amplifier must meet the following requirements:

(i) No amplifier shall be capable of achieving designed output (or designed d.c. input) power when driven with less than 50 watts mean radio frequency input power;

(ii) No amplifier shall be capable of amplifying the input RF driving signal by more than 15 decibels. (This gain limitation is determined by the ratio of the input RF driving signal to the RF output power of the amplifier where both signals are expressed in peak envelope power or mean power.) If the amplifier has a designed peak envelope power output of less than 1,500 watts, the gain allowance is reduced according-

(d) Type acceptance of external radio frequency power amplifiers or amplifier kits may be denied when denial serves the public interest, convenience or necessity by preventing the use of these amplifiers in services other than the Amateur Radio Service. Other uses of these amplifiers, such as in the Citizens Band Radio Service, are prohibited (Section 95.509). Examples of features which may result in dismissal or denial of an application for type acceptance of an external RF power amplifier include, but are not limited to, the following:

(1) Any accessible wiring which, when altered, would permit operation of the amplifier in a manner contrary to the FCC's Rules;

(2) Circuit boards or similar circuitry to facilitate the addition of components to change ly. For example, an amplifier with a designed peak envelope output power of 500 watts shall not be capable of amplifying the input RF driving signal by more than 10 decibels.

(iii) The amplifier shall not exhibit more gain than permitted by paragraph (d)(6)(ii) of this section when driven by a radio frequency input signal of less than 50 watts mean power; and

(iv) The amplifier shall be capable of sustained operation at its designed power level.

(7) Any attenuation in the input of the amplifier which, when removed or modified, would permit the amplifier to function at its designed output power when driven by a radio frequency input signal of less than 50 watts mean power.

# Subpart D — Operating Requirements and Procedures

**§ 97.78 Practice to be observed by all licensees.**

In all respects not specifically covered by these regulations each amateur station shall be operated in accordance with good engineering and good amateur practice.

**§ 97.79 Control operator requirements.**

(a) The licensee of an amateur station shall be responsible for its proper operation.

(b) Every amateur radio station, when in operation, shall have a control operator. The control operator shall be present at a control point of the station, except when the station is operated under automatic control. (Automatic control is only permitted where specifically authorized by the rules of this part.) The control operator may be the station licensee, if a licensed amateur radio operator, or may be another amateur radio operator with the required class of license and designated by the station licensee. The control operator shall also be responsible, together with the station licensee, for the proper operation of the station. (For purposes of enforcement of the rules of this part, the FCC will presume that the station licensee is, at all times, the control operator of the station, unless documentation exists to the contrary.)

(c) an amateur station may only be operated in the manner and to the extent permitted by the operator privileges authorized for the class of license held by the control operator, but may exceed those of the station licensee provided proper station identification procedures are performed.

(d) The licensee of an amateur radio station may permit any third party to participate in amateur radio communication from his station, provided that a control operator is present and continuously monitors and supervises the radio communication to insure compliance with the rules.

### § 97.83 Availability of station license.

The original license of each amateur station or a photocopy thereof shall be posted in a conspicuous place in the room occupied by the licensed operator while the station is being operated at a fixed location or shall be kept in his or her personal possession. When the station is operated at other than a fixed location, the original station license or a photocopy thereof shall be kept in the personal possession of the station licensee (or a licensed representative) who shall be present at the station while it is being operated as a portable or mobile station. The original station license shall be available for inspection by any authorized Government official at all times while the station is being operated and at other times upon request made by an authorized representative of the Commission, except when such license has been filed with application for modification or renewal thereof, or has been mutilated, lost, or destroyed, and request has been made for a duplicate license in accordance with §97.57.

### § 97.84 Station identification.

(a) Each amateur radio station shall give its call sign at the end of each communication, and every ten minutes or less during a communication.

## § 97.81 Authorized apparatus.

(a) An amateur station license authorizes the use, under control of the licensee, of all transmitting apparatus at the fixed location specified in the station license which is operated on any frequency or frequencies allocated to the Amateur Radio Service, and, in addition, authorizes the use, under control of the licensee, of portable and mobile transmitting apparatus operated at other locations.

(b) The apparatus authorized for use by paragraph (a) of this section shall be available for inspection upon request by an authorized Commission representative.

## § 97.82 Availability of operator license.

Each amateur radio operator must have the original or photocopy of his or her operator license in his or her personal possession when serving as the control operator of an amateur radio staton. The original license shall be available for inspection by any authorized government official upon request made by an authorized representative of the Commission, except when such license has been filed with application for modification or renewal thereof, or has been mutilated, lost or destroyed, and request has been made for a duplicate license in accordance with Section 97.57.

(b) Under conditions when the control operator is other than the station licensee, the station identification shall be the assigned call sign for that station. However, when a station is operated within the privileges of the operator's class of license but which exceeds those of the station licensee, station identification shall be made by following the station call sign with the operator's primary station call sign (i.e., WN4XYZ/W4XX).

(c) An amateur radio station in repeater operation or a station in auxiliary operation used to relay automatically the signals of other stations in a system of stations shall be identified by radiotelephony or radiotelegraphy at a level of modulation sufficient to be intelligible through the repeated transmission at intervals not to exceed ten minutes.

(d) When an amateur radio station is in repeater, auxiliary or beacon operation, the following additional requirements shall apply:

(1) When identifying by radiotelephony, a station in repeater operation shall transmit the word "repeater" at the end of the station call sign. When identifying by radiotelegraphy, a station in repeater operation shall transmit the fraction bar $\overline{DN}$ followed by the letters "RPT' or "R" at the end of the station call sign. (The requirements of this subparagraph do not apply to stations having call signs prefixed by the letters "WR.")

(2) When identifying by radiotelephony, a station in auxiliary operation shall transmit the word "auxiliary" at the end of the station call sign. When identifying by radiotelegraphy, a station in auxiliary operation shall transmit the fraction bar $\overline{DN}$ followed by the letters "AUX" or "A" at the end of the station call sign.

(3) When identifying by radiotelephony, a station in beacon operation shall transmit the word "beacon" at the end of the station call sign. When identifying by radiotelegraphy, a station in beacon operation shall transmit the fraction bar $\overline{DN}$ followed by the letters "BCN" or "B" at the end of the station call sign. This station identification shall be made at intervals not to exceed one minute during any period of operation.

(e) A station in auxiliary operation may be identified by the call sign of its associated station.

(f) When operating under the temporary operating authority permitted by §97.35 with privileges which exceed the privileges for the class of operator license currently held by the licensee, a licensee must identify in the following manner:

(1) On radiotelephony, by the transmission of the station call sign, followed by the word "temporary," followed by the identifier code for the new class of license for which the licensee has qualified (see §97.35).

(2) On radiotelegraphy, by the transmission

(h) At the end of an exchange of third-party communications with a station located in a foreign country, each amateur radio station shall also give the call sign of the station with which third-party communications were exchanged.

**§ 97.85 Repeater operation.**

(a) Emissions from a station in repeater operation shall be discontinued within five seconds after cessation of radiocommunications by the user station. Provisions to limit automatically the access to a station in repeater operation may be incorporated but are not mandatory.

(b) Except for operation under automatic control, as provided in paragraph (e) of this section, the transmitting and receiving frequencies used by a station in repeater operation shall be continuously monitored by a control operator immediately before and during periods of operation.

(c) A station in repeater operation shall not concurrently retransmit amateur radio signals on more than one frequency in the same amateur frequency band, from the same location.

(d) A station in repeater operation shall be operated in a manner ensuring that it is not used for one-way communications, except as provided in §97.91.

(e) A station in repeater operation, either locally controlled or remotely controlled, may also

of the station call sign, followed by the fraction bar $\overline{DN}$, followed by the identifier code for the new class of license for which the licensee has qualified (see §97.35).

(g) The identification required by this section shall be given on each frequency being utilized for transmission and shall be made in one of the following manners:

(1) By telegraphy using the international Morse code (if this identification is made by an automatic device used only for identification, the code speed shall not exceed 20 words per minute);

(2) By telephony using the English language (the Commission encourages the use of a nationally or internationally recognized standard phonetic alphabet as an aid for correct telephone identification);

(3) By telegraphy using any code authorized by §97.69(b), when the particular code is used for transmission of all or part of the communication or when the communication is transmitted in any digital code on frequencies above 50 MHz; or

(4) By video using readily legible characters when A5 emissions are used, the monochrome portions of which conform, at a minimum, to the monochrome transmission standards of §73.682(a)(6) through §73.682(a)(13), inclusive (with the exception of §73.682(a)(9)(iii) and §73.682(a)(9)(iv)).

be operated by automatic control when devices have been installed and procedures have been implemented to ensure compliance with the rules when a duty control operator is not present at a control point of the station. Upon notification by the Commission of improper operation of a station under automatic control, operation under automatic control shall be immediately discontinued until all deficiencies have been corrected.

(f) The licensee of an Amateur Radio station, before modifying an existing station in repeater operation in the National Radio Quiet Zone, or before placing his/her amateur station in repeater operation in the National Radio Quiet Zone, shall, after May 13, 1981, give written notification thereof to the Director, National Radio Astronomy Observatory, P.O. Box No. 2, Green Bank, West Virginia 24944. Station modification is any change in frequency, power, antenna height or directivity or the location of the station.

(1) The notification shall include the geographical coordinates of the antenna, antenna height, antenna directivity, if any, proposed frequency, type of emission and power.

(2) The National Radio Quiet Zone is the area bounded by 39° 15′ N. on the north, 78° 30′ W. on the east, 37° 30′ N. on the south and 80° 30′ W. on the west.

(3) If an objection to the proposed operation is received by the Commission from the National Radio Astronomy Observatory at Green Bank,

Pocahontas County, West Virginia, for itself or on behalf of the Naval Research Laboratory at Sugar Grove, Pendleton County, West Virginia, within 20 days from the date of notification, the Commission will consider all aspects of the problem and take whatever action is deemed appropriate.

(g) Each station in repeater operation transmitting with an effective radiated power greater than 100 watts on frequencies between 29.5 and 420 MHz, or 400 watts on frequencies between 420 and 1215 MHz, shall have the following information included in the station records during any period of operation:

(1) The location of the station transmitting antenna marked upon a topographic map having contour intervals and having a scale of 1:250,00 (indexes and ordering information for suitable maps are available from the U.S. Geological Survey, Washington, D.C. 20242, or from the Federal Center, Denver, CO 80255);

(2) The transmitting antenna height above average terrain (see Appendix 5);

(3) The effective radiated power in the horizontal plane for the main lobe of antenna pattern, calculated for the maximum transmitter output power which occurs during operation;

(4) The maximum transmitter output power which occurs during operation;

(5) The loss in the transmission line between have been installed and procedures have been implemented to ensure compliance with the rules when the duty control operator is not present at a control point of the station.

(c) Beacon operation shall cease upon notification by an Engineer-in-Charge of a Commission field facility that the station is operating improperly or causing undue interference to other operations. Beacon operation shall not resume without prior approval of the Engineer-in-Charge.

(d) The licensee of an amateur radio station, before modifying an existing station in automatically-controlled beacon operation in the National Radio Quiet Zone, or before placing his/her station in the National Radio Quiet Zone, shall give written notification thereof to the Director, National Radio Astronomy Observatory, P.O. Box 2, Green Bank, West Virginia 24944. Station modification is any change in frequency, power, antenna height or directivity, or the location of the station. In such cases, the rules of §97.85(f)(1), (2) and (3) shall apply.

### § 97.88 Operation of a station by remote control.

An amateur radio station may be operated by remote control only if there is compliance with the following:

(a) A photocopy of the license for the remotely

the transmitter and the antenna (including devices such as duplexers, cavities or circulators), expressed in decibels; and

(6) the relative gain in the horizontal plane of the transmitting antenna.

**§ 97.86 Auxiliary operation.**

(a) A station in auxiliary operation, either locally controlled or remotely controlled, may also be operated by automatic control when it is operated as part of a system of stations in repeater operation operated under automatic control.

(b) If a station in auxiliary operation is relaying signals of another amateur radio station(s) to a station in repeater operation, the station in auxiliary operation may use an input (receiving) frequency in frequency bands reserved for auxiliary operation, repeater operation, or both.

(c) A station in auxiliary operation shall be used only to communicate with stations shown in the system network diagram.

**§ 97.87 Beacon operation.**

(a) A station in beacon operation shall not concurrently operate on more than one frequency in the same amateur frequency band, from the same location.

(b) A station in beacon operation, either locally controlled or remotely controlled, may also be operated by automatic control when devices controlled station shall be posted in a conspicuous place at the station location.

(b) The name, address, and telephone number of the remotely controlled station licensee and at least one control operator shall be posted in a conspicuous place at the remotely controlled transmitter location.

(c) Except for operation under automatic control, a control operator shall be on duty when the station is being remotely controlled. Immediately before and during the periods the remotely controlled station is in operation, the frequencies used for emission by the remotely controlled station shall be monitored by the control operator. The control operator shall terminate all transmissions upon any deviation from the rules.

(d) Provisions must be incorporated to limit transmission to a period of no more than 3 minutes in the event of malfunction in the control link.

(e) A station in repeater operation shall be operated by radio remote control only when the control link uses frequencies other than the input (receiving) frequencies of the station in repeater operation.

(f) The station records shall include during any period of operation:

(1) The names, addresses, and call signs of all persons authorized by the station licensee to be control operators; and

(2) A functional block diagram of the control link and a technical explanation sufficient to describe its operation.

(g) Each remotely controlled station shall be protected against unauthorized station operation, whether caused by activation of the control link, or otherwise.

### § 97.89 Point of Communications.

(a) Amateur stations may communicate with:

(1) Other amateur stations, excepting those prohibited by Appendix 2.

(2) Stations in other services licensed by the Commission and with the U.S. Government stations for civil defense purposes in accordance with Subpart F of this Part, in emergencies and, on a temporary basis, for test purposes.

(3) Any station which is authorized by the Commission to communicate with amateur stations.

(b) Amateur radio stations may transmit one-way signals to receiving apparatus while in beacon operation or radio control operation.

### § 97.90 System network diagram required.

When a station has one or more associated stations, that is, stations in repeater or auxiliary operation, a system network diagram (see

### § 97.93 Modulation of carrier.

Except for brief tests or adjustments, an amateur radiotelephone station shall not emit a carrier wave on frequencies below 51 megahertz unless modulated for the purpose of communication. Single audio frequency tones may be transmitted for test purposes of short duration for the development and perfection of amateur radio telephone equipment.

## STATION OPERATION AWAY FROM AUTHORIZED LOCATION

### § 97.95 Operation away from the authorized fixed station location.

(a) Operation within the United States, its territories or possessions is permitted as follows:

(1) When there is no change in the authorized fixed operation station location, an amateur radio station, other than a military recreation station, may be operated portable or mobile under its station license anywhere in the United States, its territories or possessions, subject to §97.61.

(2) When the authorized fixed station location is changed, the licensee shall submit an application for modification of the station license in accordance with §97.47.

(b) When outside the continental limits of the United States, its territories, or possessions, an

§97.3(v)) shall be included in the station records during any period of operation.

**§ 97.91 One-way communications.**

In addition to beacon operation and radio control operation, the following kinds of one-way communications, addressed to amateur stations, are authorized and will not be construed as broadcasting: (a) Emergency communications, including *bona fide* emergency drill practice transmissions; (b) Information bulletins consisting solely of subject matter having direct interest to the amateur radio service as such; (c) Round-table discussions or net-type operations where more than two amateur stations are in communication, each station taking a turn at transmitting to other station(s) of the group; and (d) Code practice transmissions intended for persons learning or improving proficiency in the international Morse code.

**§ 97.92 Record of operations.**

When deemed necessary by the Engineer-in-Charge (EIC) of a Commission field facility to assure compliance with the rules of this Part, a station licensee shall maintain a record of station operations containing such items of information as the EIC may require under Section 0.314(x).

amateur radio station may be operated as portable or mobile only under the following conditions:

(1) Operation may not be conducted within the jurisdiction of a foreign government except pursuant to, and in accordance with express authority granted to the licensee by such foreign government. When a foreign government permits Commission licensees to operate within its territory, the amateur frequency bands which may be used shall be as prescribed or limited by that government. (See Appendix 4 of this Part for the text of treaties or agreements between the United States and foreign governments relative to reciprocal amateur radio operation.)

(2) When outside the jurisdiction of a foreign government, amateur operation may be conducted within ITU Region 2 subject to the limitations of, and on those frequency bands listed in §97.61.

(3) When outside the jurisdiction of a foreign government, amateur operation may be conducted within ITU Regions 1 and 3 on the following frequencies, subject to the limitations and provisions of Section IV of Article 5 of the Radio Regulations of the ITU:

(i)

| | |
|---|---|
| REGION 1 | 28.0-29.7 MHz |
| 3.5-3.8 MHz | 144-146 MHz |
| 7.0-7.1 MHz | 430-440 MHz |
| 14.0-14.35 MHz | 1215-1300 MHz |
| 21.0-21.45 MHz | 2300-2450 MHz |
| REGION 3 | 28.0-29.7 MHz |
| 1.8-2.0 MHz | 50.0-54.0 MHz |
| 3.5-3.9 MHz | 144-148 MHz |
| 7.0-7.1 MHz | 420-450 MHz |
| 14.0-14.35 MHz | 1215-1300 MHz |
| 21.0-21.45 MHz | 2300-2450 MHz |

(ii) Operation on amateur frequency bands above 2450 MHz may be conducted subject to the limitations and provisions of Section IV of Article 5 of the Radio Regulations of the ITU.

(4) Except as otherwise provided, amateur operation conducted outside the jurisdiction of a foreign government shall comply with all requirements of Part 97 of this chapter.

CHART OF REGIONS AS DEFINED IN TABLE OF FREQUENCY ALLOCATIONS

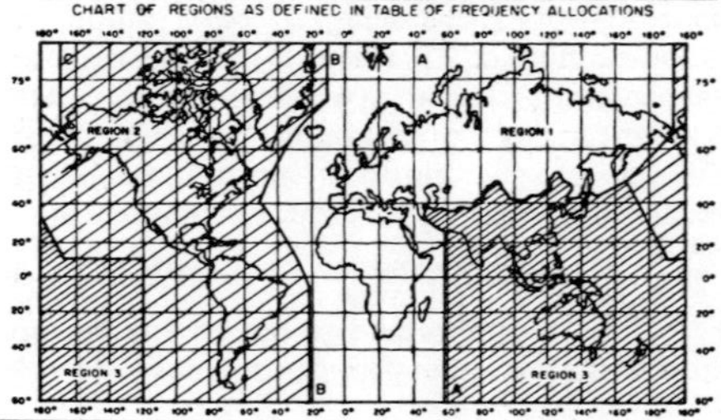


tion call sign and the licensee's name and address is affixed to the transmitter.

(a) Station identification is not required for transmission directed only to a remote model craft or vehicle.

(b) Transmissions containing only control signals directed only to a remote model craft or vehicle are not considered to be codes or ciphers in the context of the meaning of §97.117.

### § 97.101 Mobile stations aboard ships or aircraft.

In addition to complying with all other applicable rules, an amateur mobile station operated on board a ship or aircraft must comply with all of the following special conditions: (a) The installation and operation of the amateur mobile station shall be approved by the master of the ship or captain of the aircraft; (b) The amateur mobile station shall be separate from and independent of all other radio equipment, if any, installed on board the same ship or aircraft; (c) The electrical installation of the amateur mobile station shall be in accord with the rules applicable to ships or aircraft as promulgated by the appropriate government agency; (d) The operation of the amateur mobile station shall not interfere with the efficient operation of any other radio equipment installed on board the same ship or aircraft; and (e) The amateur mobile station and its associated equipment, either in itself or in its method of operation, shall not constitute a hazard to the safety of life or property.

Note: Region 2 is defined as follows: On the east, a line (B) extending from the North Pole along meridian 10° of Greenwich to its intersection with parallel 72°; thence by Great Circle Arc to the intersection of meridian 50° W and parallel 40° N; thence by Great Circle Arc to the intersection of meridian 20° W and parallel 10° S; thence along meridian 20 degrees west to the South Pole. On the west, a line (C) extending from the North Pole by Great Circle Arc to the intersection of parallel 65°, 30′ N with the international boundary in Bering Strait; thence by Great Circle Arc to the intersection of meridian 165° East of Greenwich and parallel 50° N; thence by Great Circle Arc to the intersection of meridian 170° West and parallel 10° N; thence along parallel 10° N to its intersection with meridian 120° West thence along meridian 12° West to the South Pole.

## SPECIAL PROVISIONS

### § 97.99 Stations used only for radio control of remote model craft and vehicles.

An amateur radio station in radio control operation with a mean output power not exceeding one watt may, when used for the control of a remote model craft or vehicle, be operated under the special provisions of this section, provided that a writing indicating the sta-

## EMERGENCY OPERATIONS

### § 97.107 Operation in emergencies.

In the event of an emergency disrupting normally available communication facilities in any widespread area or areas, the Commission, in its discretion, may declare that a general state of communications emergency exists, designate the area or areas concerned, and specify the amateur frequency bands, or segments of such bands, for use only by amateurs participating in emergency communication within or with such affected area or areas. Amateurs desiring to request the declaration of such a state of emergency should communicate with the Commission's Engineer-in-Charge of the area concerned. Whenever such declaration has been made, operation of and with amateur stations in the area concerned shall be only in accordance with the requirements set forth in this section, but such requirements shall in nowise affect other normal amateur communication in the affected areas when conducted on frequencies not designated for emergency operation.

(a) All transmissions within all designated

amateur communications bands[1] other than communications relating directly to relief work, emergency service or the establishment and maintenance of efficient Amateur Radio networks for the handling of such communications shall be suspended. Incidental calling, answering, testing or working (including casual conversations, remarks or messages) not pertinent to constructive handling of the emergency situation shall be prohibited within these bands.

(b) The Commission may designate certain amateur stations to assist in the promulgation of information relating to the declaration of a general state of communications emergency, to monitor the designated amateur emergency communications bands, and to warn non-complying stations observed to be operating in those bands. Such station, when so designated, may transmit for that purpose on any frequency or frequencies authorized to be used by that station, provided such transmissions do not interfere with essential emergency communications in progress; however, such transmissions shall preferably be made on authorized frequencies immediately adjacent to those segments of the amateur bands being cleared for the emergency. Individual transmissions for the purpose of advising other stations of the existence of the communications emergency shall refer to this section by number (§97.107) and shall specify, briefly and concisely, the date of the Commission's declaration, the area and nature of the emergency, and the amateur frequency bands or segments of such bands which constitute the amateur emergency communications bands at the time. The designated stations shall not enter into discussions with other stations beyond furnishing essential facts relative to the emergency, or acting as advisors to stations desiring to assist in the emergency, and the operators of such designated stations shall report fully to the Commission the identity of any stations failing to comply, after notice, with any of the pertinent provisions of this section.

(c) The special conditions imposed under the provisions of this section shall cease to apply only after the Commission or its authorized representative, shall have declared such general state of communications emergency to be terminated; however, nothing in this paragraph shall be deemed to prevent the Commission from modifying the terms of its declaration from time to time as may be necessary during the period of

[1]The frequency 5167.5 kHz may be used by any station authorized under this part to communicate with any other station in the State of Alaska for emergency communications. No airborne operations will be permitted on this frequency. All stations operating on this frequency must be located in or within 50

nautical miles of the State of Alaska. The frequency 5167.5 kHz may be used by licensees in the Alaska-private fixed service for calling and listening, but only for establishing communication before switching to another frequency.

a communications emergency, or from removing those conditions with respect to any amateur frequency band or segment of such band which no longer appears essential to the conduct of the emergency communications.

# Subpart E — Prohibited Practices and Administrative Sanctions

**PROHIBITED TRANSMISSIONS AND PRACTICES**

**§ 97.110 Business communications prohibited.**

The transmission of business communications by an amateur radio station is prohibited, except for emergency communications as defined in this Part.

**§ 97.111 Limitations on international communications.**

Transmissions between amateur radio stations of different countries, when permitted, must be limited to messages of a technical nature relating to tests, and, to remarks of a personal character for which, by reason of their unimportance, recourse to the public telecommunications service is not justified.

**§ 97.112 No remuneration for use of station.**

(a) An amateur station shall not be used to transmit or receive messages for hire, nor for communication for material compensation, direct or indirect, paid or promised.

(b) Control operators of a club station may be compensated when the club station is operated primarily for the purpose of conducting amateur radiocommunication to provide telegraphy practice transmissions intended for persons learning or improving proficiency in the international Morse code, or to disseminate information bulletins consisting solely of subject matter having direct interest to the Amateur Radio Service provided:

(1) The station conducts telegraphy practice and bulletin transmission for at least 40 hours per week.

(2) The station schedules operations on all allocated medium and high frequency amateur bands using reasonable measures to maximize coverage.

(3) The schedule of normal operating times and frequencies is published at least 30 days in advance of the actual transmissions.

Control operators may accept compensation only for such periods of time during which the station is transmitting telegraphy practice or bulletins. A control operator shall not accept any direct or indirect compensation for periods during which the station is transmitting material other than telegraphy practice or bulletins.

**§ 97.113 Broadcasting prohibited.**

Subject to the provisions of §97.91, an amateur station shall not be used to engage in any form of broadcasting, that is, the dissemination of radio communications intended to be received by the public directly or by the intermediary of relay stations, nor for the retransmission by automatic means of programs or signals emanating from any class of station other than amateur. The foregoing provisions shall not be construed to prohibit amateur operators from giving their consent to the rebroadcast by broadcast stations of the

**§ 97.117 Codes and ciphers prohibited.**

The transmission by radio of messages in codes or ciphers in domestic and international communications to or between amateur stations is prohibited. All communications regardless of type of emission employed shall be in plain language except that generally recognized abbreviations established by regulation or custom and usage are permissible as are any other abbreviations or signals where the intent is not to obscure the meaning but only to facilitate communications.

**§ 97.119 Obscenity, indecency, profanity.**

No licensed radio operator or other person shall transmit communications containing obscene, indecent, or profane words, language, or meaning.

**§ 97.121 False signals.**

No licensed radio operator shall transmit false or deceptive signals or communications by radio, or any call letter or signal which has not been assigned by proper authority to the radio station he is operating.

**§ 97.123 Unidentified communications.**

No licensed radio operator shall transmit unidentified radio communications or signals.

transmissions of their amateur stations, provided that the transmissions of the amateur station shall not contain any direct or indirect reference to the rebroadcast.

**§ 97.114 Third-party traffic.**

The transmission or delivery of the following amateur radiocommunications is prohibited:

(a) International third-party traffic except with countries which have assented thereto.

(b) Third-party traffic involving material compensation, either tangible or intangible, direct or indirect, to a third-party, a station licensee, a control operator, or any other person.

(c) Except for an emergency communication as defined in this Part, third-party traffic consisting of business communications on behalf of any party.

**§ 97.115 Music prohibited.**

The transmission of music by an amateur station is forbidden.

**§ 97.116 Amateur radiocommunication for unlawful purposes prohibited.**

The transmission of radio communication or messages by an amateur radio station for any purpose, or in connection with any activity, which is contrary to Federal, State or local law is prohibited.

**§ 97.125 Interference.**

No licensed radio operator shall willfully or maliciously interfere with or cause interference to any radio communication or signal.

**§ 97.126 Retransmitting radio signals.**

(a) An amateur radio station, except a station in repeater operation or auxiliary operation, shall not automatically retransmit the radio signals of other amateur radio stations.

(b) A remotely controlled station, other than a remotely controlled station in repeater operation or auxiliary operation, shall automatically retransmit only the radio signals of stations in auxiliary operation shown on the remotely controlled station's system network diagram.

**§ 97.127 Damage to apparatus.**

No licensed radio operator shall willfully damage, or cause or permit to be damaged, any radio apparatus or installation in any licensed radio station.

**§97.129 Fraudulent licenses.**

No licensed radio operator or other person shall obtain or attempt to obtain, or assist another to obtain or attempt to obtain, an operator license by fraudulent means.

## ADMINISTRATIVE SANCTIONS

### § 97.131 Restricted operation.

(a) If the operation of an amateur station causes general interference to the reception of transmissions from stations operating in the domestic broadcast service when receivers of good engineering design including adequate selectivity characteristics are used to receive such transmission and this fact is made known to the amateur station licensee, the amateur station shall not be operated during the hours from 8 p.m. to 10:30 p.m., local time, and on Sunday for the additional period from 10:30 a.m. until 1 p.m., local time, upon the frequency or frequencies used when the interference is created.

(b) In general, such steps as may be necessary to minimize interference to stations operating in other services may be required after investigation by the Commission.

### § 97.133 Second notice of same violation.

In every case where an amateur station licensee is cited within a period of 12 consecutive months for the third violation of §§97.61, 97.63, 97.65, or 97.73, the station licensee, if directed by the Commission, shall not operate the station and shall not permit it to be operated from 8 a.m. to 12 midnight, local time, except for the purpose of transmitting a prearranged test to be observed by a monitoring station of the Commission to be designated in each particular case. The station shall not be permitted to resume operation during these hours until the licensee is authorized by the Commission following the test, to resume full-time operation. The results of the test and the licensee's record shall be considered in determining the advisability of suspending the operator license or revoking the station license, or both.

### § 97.137 Answers to notices of violation.

Any licensee receiving official notice of a violation of the terms of the Communications of 1934, as amended, any legislative act, Executive order, treaty to which the United States is a party, or the rules of the Federal Communica-

for the second violation of the provisions of §§97.61, 97.63, 97.65, or 97.73, the station licensee, if directed to do so by the Commission, shall not operate the station and shall not permit it to be operated from 6 p.m. to 10:30 p.m., local time, until written notice has been received authorizing the resumption of full-time operation. This notice will not be issued until the licensee has reported on the results of tests which he/she has conducted with at least two other amateur stations at hours other than 6 p.m. to 10:30 p.m., local time. Such tests are to be made for the specific purpose of aiding the licensee in determining whether the emissions of the station are in accordance with the Commission's rules. The licensee shall report to the Commission the observations made by the cooperating amateur licensee in relation to reported violations. This report shall include a statement as to the corrective measures taken to insure compliance with the rules.

**§ 97.135 Third notice of same violation.**

In every case where an amateur station licensee

tions Commission, shall, within 10 days from such receipt, send a written answer direct to the office of the Commission originating the official notice: *Provided, however,* that if an answer cannot be sent or an acknowledgement made within such 10-day period by reason of illness or other unavoidable circumstances, acknowledgement and answer shall be made at the earliest practicable date with a satisfactory explanation of the delay. The answer to each notice shall be complete in itself and shall not be abbreviated by reference to other communications or answers to other notices. If the notice relates to some violation that may be due to the physical or electrical characteristics of transmitting apparatus, the answer shall state fully what steps, if any, are taken to prevent future violations, and if any new apparatus is to be installed, the date such apparatus was ordered, the name of the manufacturer, and promised date of delivery. If the notice of violation relates to some lack of attention to or improper operation of the transmitter, the name of the operator in charge shall be given.

# Subpart F — Radio Amateur Civil Emergency Services (RACES)

**GENERAL**

**§ 97.161 Basis and purpose.**

The Radio Amateur Civil Emergency Service provides for amateur radio operation for civil defense communications purposes only, during periods of local, regional or national civil emergencies, including any emergency which may necessitate invoking of the President's War Emergency Powers under the provisions of section 606 of the Communications Act of 1934, as amended.

**§ 97.163 Definitions.**

For the purposes of this Subpart, the following definitions are applicable:

(a) *Radio Amateur Civil Emergency Service.* A radio communication service conducted by volunteer licensed amateur radio operators, for providing emergency radiocommunications to local, regional, or state civil defense organizations.

(b) *RACES station.* An amateur radio station licensed to a civil defense organization, at a local, regional, or state civil defense organization.

(b) Only modification and/or renewal station licenses will be issued for RACES stations. No new licenses will be issued for RACES stations.

**§ 97.173 Application for RACES station license.**

(a) Each application for a RACES station license shall be made on the FCC Form 610-B.

(b) The application shall be signed by the civil defense official responsible for the coordination of all civil defense activities in the area concerned.

(c) The application shall be countersigned by the responsible official for the governmental entity served by the civil defense organization.

(d) If the application is for a RACES station to be in any special manner covered by §97.42, those showings specified for non-RACES stations shall also be submitted.

**§ 97.175 Amateur radio station registration in civil defense organization.**

No amateur radio station shall be operated in

specific land location, for the purpose of providing the facilities for amateur radio operators to conduct amateur radiocommunications in the Radio Amateur Civil Emergency Service.

**§ 97.165 Applicability of rules.**

In all cases not specifically covered by the provisions contained in this Subpart, amateur radio stations and RACES stations shall be governed by the provisions of the rules governing amateur radio stations and operators (Subpart A through E of this part).

## STATION AUTHORIZATIONS

**§ 97.169 Station license required.**

No transmitting station shall be operated in the Radio Amateur Civil Emergency Service unless:

(a) The station is licensed as a RACES station by the Federal Communications Commission, or

(b) The station is an amateur station licensed by the Federal Communications Commission, and is certified by the responsible civil defense organization as registered with that organization.

**§ 97.171 Eligibility for RACES station license.**

(a) A RACES station will only be licensed to the Radio Amateur Civil Emergency Service unless it is certified as registered in a civil defense organization by that organization.

## OPERATING REQUIREMENTS

**§ 97.177 Operator requirements.**

No person shall be the control operator of a RACES station, or shall be the control operator of an amateur radio station conducting communications in the Radio Amateur Civil Emergency Service unless that person holds a valid amateur radio operator license and is certified as enrolled in a civil defense organization by that organization.

**§ 97.179 Operator privileges.**

Operator privileges in the Radio Amateur Civil Emergency Service are dependent upon, and identical to, those for the class of operator license held in the Amateur Radio Service.

**§ 97.181 Availability of RACES station license and operator licenses.**

(a) The original license of each RACES station, or a photocopy thereof, shall be attached to each transmitter of such station, and at each control point of such station. Whenever a photocopy of the RACES station license is utilized in compliance with this requirement, the original station license shall be available for inspection by any authorized Government official at all times when the station is being operated and at other times upon request made by an authorized representative of the Commission, except when such license has been filed with application for modification or renewal thereof, or has been mutiliated, lost, or destroyed, and request has been made for a duplicate license in accordance with §97.57.

(b) In addition to the operator license availability requirements of §97.82, a photocopy of the control operator's amateur radio operator license shall be posted at a conspicuous place at the control point for the RACES station.

## TECHNICAL REQUIREMENTS

### § 97.185 Frequencies available.

(a) All of the authorized frequencies and emissions allocated to the Amateur Radio Service are also available to the Radio Amateur Civil Emergency Service on a shared basis.

(b) In the event of an emergency which necessitates the invoking of the President's War Emergency Powers under the provisions of §606 of the Communications Act of 1934, as amended, unless otherwise modified or directed, RACES stations and amateur radio stations participating in RACES will be limited in operation to the following:

1240-1300 MHz band.

(4) Those stations operating in the band 220-225 MHz shall not cause harmful interference to, and must tolerate any interference from, the Government Radiolocation Service until January 1, 1990. Additionally, the Fixed and Mobile Services shall have equal right of operation.

(5) In the band 420-430 MHz, no station shall operate North of Line A. Line A begins at Aberdeen, Washington, running by great circle arc to the intersection of 48° N., 120° W., thence along parallel 48° N., to the intersection of 95° W., thence by great circle arc through the southernmost point of Duluth, Minn., thence by great circle arc to 45° N., 85° W., thence southward along meridian 85° W., to its intersection with parallel 41° N., thence along parallel 41° W., to its intersection with meridian 82° W., thence by great circle arc through the southernmost point of Bangor, Maine, thence by great circle arc through the southernmost point of Searsport, Maine, at which point it terminates.

(6) In the band 420-450 MHz and within the following areas, the peak envelope power output of a transmitter used in the Amateur Radio Service shall not exceed 50 watts, unless expressly authorized by the Commission after mutual agreement, on a case-by-case basis, between the Federal Communications Commission Engineer-in-Charge at the applicable District Office and

Frequency or Frequency Bands

| *kHz* | *Limitations* | *MHz* | *Limitations* |
|---|---|---|---|
| 1800-1825 | | 28.55-28.75 | |
| 1975-2000 | | 29.237-29.273 | |
| 3500-3550 | | 29.45-29.65 | |
| 3930-3980 | | 50.35-50.75 | |
| 3984-4000 | | 52-54 | |
| 3997 | 2 | 53.30 | 2 |
| 7079-7125 | | 53.35-53.75 | |
| 7245-7255 | | 144.50-145.71 | |
| 14047-14053 | | 146-148 | |
| 14220-14230 | | 220-225 | 4 |
| 14331-14350 | | 420-450 | 3, 5, 6 |
| 21047-21053 | | 1240-1300 | 3 |
| 21228-21267 | | 2390-2450 | 3 |

(c) Limitations:

(1) (Reserved)

(2) For use in emergency areas when required to make initial contact with a military unit; also, for communications with military stations on matters requiring coordination.

(3) Those stations operating in the bands 420-450, 1240-1300 and 2390-2450 MHz shall not cause harmful interference to, and must tolerate any interference from, the Government Radiolocation Service; and also the Aeronautical Radionavigation Service in the case of the Military Area Frequency Coordinator at the applicable military base:

(i) Those portions of Texas and New Mexico bounded on the south by latitude 31° 45′ North, on the east by longitude 104° 00′ West, on the north by latitude 34° 30′ North, and on the west by longitude 107° 30′ West;

(ii) The entire State of Florida including the Key West area and the areas enclosed within a 200-mile radius of Patrick Air Force Base, Florida (latitude 28° 21′ North, longitude 80° 43′ West), and within North, longitude 86° 30′ West);

(iii) The entire State of Arizona;

(iv) Those portions of California and Nevada south of latitude 37° 10′ North, and the areas enclosed within a 200-mile radius of the Pacific Missile Test Center, Point Mugu, California (latitude 34° 09′ North, longitude 119° 11′ West).

(v) In the State of Massachusetts within a 160-kilometer (100 miles) radius around locations at Otis Air Force Base, Massachusetts (latitude 41° 45′ North, longitude 70° 32′ West).

(vi) In the State of California within a 240-kilometer (150 mile) radius around locations at Beale Air Force Base, California (latitude 39° 08′ North, longitude 121° 26′ West).

(vii) In the State of Alaska within a

160-kilometer (100 mile) radius of Clear, Alaska (latitude 64° 17′ North, longitude 149° 10′ West). (The Military Area Frequency Coordinator for this area is located at Elmendorf Air Force Base, Alaska.)

(viii) In the State of North Dakota (latitude 48° 43′ North, longitude 97° 54′ West). (The Military Area Frequency Coordinator for this area can be contacted at: HQ SAC/SXOE, Offutt Air Force Base, Nebraska 68113.)

**§ 97.189 Point of communications.**

(a) RACES stations may only be used to communicate with:

(1) Other RACES stations;

(2) Amateur radio stations certified as being registered with a civil defense organization, by that organization;

(3) Stations in the Disaster Communications Service;

(4) Stations of the Unites States Government authorized by the responsible agency to exchange communications with RACES stations;

(5) Any other station in any other service regulated by the Federal Communications Commission, whenever such station is authorized by the Commission, to exchange communications with stations in the Radio Amateur Civil Emergency Service.

(b) Amateur radio stations registered with a civil defense organization may only be used to communicate with:

mission, whenever such station is authorized by the Commission to exchange communications with stations in the Radio Amateur Civil Emergency Service.

**§ 97.191 Permissible communications.**

All communications in the Radio Amateur Civil Emergency Service must be specifically authorized by the civil defense organization for the area served. Stations in this service may transmit only civil defense communications of the following types:

(a) Communications concerning impending or actual conditions jeopardizing the public safety, or affecting the national defense or security during periods of local, regional, or national civil emergencies:

(1) Communications directly concerning the immediate safety of life or individuals, the immediate protection of property, maintenance of law and order, alleviation of human suffering and need, and the combating of armed attack or sabotage;

(2) Communications directly concerning the accumulation and dissemination of public information or instructions to the civilian population essential to the activities of the civil defense organization or other authorized governmental or relief agencies.

(b) Communications for training drills and tests necessary to ensure the establishment and maintenance of orderly and efficient operation of the Radio Amateur Civil Emergency Service

(1) RACES stations licensed to the civil defense organization with which the amateur radio station is registered:

(2) Any of the following stations upon authorization of the responsible civil defense official for the organization in which the amateur radio station is registered:

(i) Any RACES station licensed to other civil defense organizations;

(ii) Amateur radio stations registered with the same or another civil defense organization;

(iii) Stations in the Disaster Communications Service;

(iv) Stations of the United States Government authorized by the responsible agency to exchange communications with RACES stations;

(v) Any other station in any other service regulated by the Federal Communications Commission as ordered by the responsible civil defense organization served. Such tests and drills may not exceed a total time of one hour per week.

(c) Brief one way transmissions for the testing and adjustment of equipment.

**§ 97.193 Limitations on the use of RACES stations.**

(a) No station in the Radio Amateur Civil Emergency Service shall be used to transmit or to receive messages for hire, nor for communications for material compensation, direct or indirect, paid or promised.

(b) All messages which are transmitted in connection with drills or tests shall be clearly identified as such by use of the words "drill" or "test", as appropriate, in the body of the messages.

# Subpart G — Operation of Amateur Radio Stations in the United States by Aliens Pursuant to Reciprocal Agreements

**§ 97.301 Basis, purpose, and scope.**

(a) The rules in this subpart are based on, and are applicable solely to, alien amateur operations pursuant to section 303(1)(3) and 310(a) of the Communications Act of 1934, as amended. (See Pub. L 93-505, 88 Stat. 1576.)

(b) The purpose of this subpart is to implement Public Law 88-383 by prescribing the rules under which an alien, who holds an amateur operator and station license issued by his government (referred to in this subpart as an alien amateur), may operate an amateur radio station in the United States, in its possessions, and the Commonwealth of Puerto Rico (referred to in this subpart only as the United States).

**§ 97.303 Permit required.**

Before he may operate an amateur radio station in the United States, under the provisions of sections 303(1)(3) and 310(c) of the Communications Act of 1934, as amended, an alien amateur licensee must obtain a permit for such operation from the Federal Communications Commission. A permit for such operation shall be issued only to an alien holding a valid amateur operator and station authorization from his government, and only when there is in effect a bilateral agreement between the United States and that government for such operation on a reciprocal basis by United States amateur radio operators.

**§ 97.305 Application for permit.**

(a) Application for a permit shall be made on FCC Form 610-A. Form 610-A may be obtained from the Commission's Washington, DC, office, from any of the Commission's field offices and, in some instances, from United States missions abroad.

(b) The application form shall be completed in full in English and signed by the applicant. A photocopy of the applicant's amateur operator and station license issued by his government shall request the Commission to reconsider its action.

(c) Normally, a permit will be issued to expire 1 year after issuance but in no event after the expiration of the license issued to the alien amateur by his government.

**§ 97.309 Modification, suspension, or cancellation of permit.**

At any time the Commission may, in its discretion, modify, suspend, or cancel any permit issued under this subpart. In this event, the permittee will be notified of the Commission's action by letter mailed to his mailing address in the United States and the permittee shall comply immediately. A permittee may, within 90 days of the mailing of such letter, request the Commission to reconsider its action. The filing of a request for reconsideration shall not stay the effectiveness of that action, but the Commission may stay its action on its own motion.

**§ 97.311 Operating conditions.**

(a) The alien amateur may not under any circumstances begin operation until he has received a permit issued by the Commission.

(b) Operation of an amateur station by an alien amateur under a permit issued by the Commission must comply with all of the following:

(1) The terms of the bilateral agreement between the alien amateur's goverment and the

be filed with the application. The Commission may require the applicant to furnish additional information. The application must be filed by mail or in person with the Federal Communications Commission, Gettysburg, PA 17325, USA. To allow sufficient time for processing, the application should be filed at least 60 days before the date on which the applicant desires to commence operation.

### § 97.307 Issuance of permit.

(a) The Commission may issue a permit to an alien amateur under such terms and conditions as it deems appropriate. If a change in the terms of a permit is desired, an application for modification of the permit is required. If operation beyond the expiration date of a permit is desired, an application for renewal of the permit is required. In any case in which the permittee has, in accordance with the provisions of this subpart, made a timely and sufficient application for renewal of an unexpired permit, such permit shall not expire until the application has been finally determined. Applications for modification or for renewal of a permit shall be filed on FCC Form 610-A.

(b) The Commission, in its discretion may deny any application for a permit under this subpart. If an application is denied, the applicant will be notified by letter. The applicant may, within 90 days of the mailing of such letter, re-

government of the United States;

(2) The provisions of this subpart and of Subparts A through E of this part;

(3) The operating terms and conditions of the license issued to the alien amateur by his government; and

(4) Any further conditions specified on the permit issued by the Commission.

### § 97.313 Station identification.

(a) The alien amateur shall identify his station as follows:

(1) Radio telegraph operation: The amateur shall transmit the call sign issued to him by the licensing country followed by a slant (/) sign and the United States amateur call sign prefix letter(s) and number appropriate to the location of his station.

(2) Radiotelephone operation: The amateur shall transmit the call sign issued to him by the licensing country followed by the words "fixed", "portable" or "mobile", as appropriate, and the United States amateur call sign prefix letter(s) and number appropriate to the location of his station. The identification shall be made in the English language.

(b) At least once during each contact with another amateur station, the alien amateur shall indicate, in English, the geographical location of his station as nearly as possible by city and state, commonwealth or possession.

# Subpart H — Amateur-Satellite Service

## GENERAL

**§ 97.401 Purposes.**

The Amateur-Satellite Service is a radiocommunication service using stations on earth statellites for the same purpose as those of the Amateur Radio Service.

**§ 97.403 Definitions.**

(a) *Space operation*. Space-to-earth, and space-to-space, Amateur Radio communication from a station which is beyond, is intended to go beyond, or has been beyond the major portion of the earth's atmosphere.

(b) *Earth operation*. Earth-to-space-to-earth amateur radiocommunication by means of radio signals automatically retransmitted by stations in space operation.

(c) *Telecommand operation*. Earth-to-space Amateur Radio communication to initiate, modify, or terminate functions of a station in space operation.

(d) *Telemetry*. Space-to-earth transmissions, by a station in space operation, of results of measurements made in the station, including those relating to the function of the station.

license (see §97.7).

**§ 97.413 Space operations requirements.**

An Amateur Radio station may be in space operation where:

(a) The station has not been ordered by the Commission to cease radio transmissions.

(b) The station is capable of effecting a cessation of radio transmissions by commands transmitted by station(s) in telecommand operation whenever such cessation is ordered by the Commission.

(c) There are, in place, sufficient Amateur Radio stations licensed by the Commission capable of telecommand operation to effect cessation of space operation, whenever such is ordered by the Commission.

(d) The notification required by §97.423 are on file with the Commission.

## TECHNICAL REQUIREMENTS

**§ 97.415 Frequencies available.**

The following frequency bands are available for space operation, earth operation, and telecommand operation:

**§ 97.405 Applicability of rules.**

The rules contained in this Subpart apply to radio stations in the Amateur-Satellite Service. All cases not specifically covered by the provisions of this Subpart shall be governed by the provisions of the rules governing Amateur Radio stations and operators (Subpart A through E of this Part).

**§ 97.407 Eligibility for space operation.**

Amateur Radio stations licensed to Amateur Extra Class operators are eligible for space operation (see §97.403(a)). The station licensee may permit any Amateur Radio operator to be the control operator, subject to the privileges of the control operator's class of license (see §97.7).

**§ 97.409 Eligibility for earth operation.**

Any Amateur Radio station is eligible for earth operation (see §97.403(b)), subject to the privileges of the control operator's class of license (see §97.7).

**§ 97.411 Eligibility for telecommand operation.**

Any Amateur Radio station designated by the licensee of a station in space operation is eligible to conduct telecommand operation with the station in space operation, subject to the privileges of the control operator's class of

**Frequency bands**

| *kHz* | *MHz* | *GHz* |
|---|---|---|
| 7000-7100 | 21.00-21.45 | 24-24.05 |
| 14,000-14,250 | 28.00-29.70 | |
| | 144-146 | |
| | 435-438[1] | |

## SPECIAL PROVISIONS

**§ 97.417 Space operation.**

(a) Stations in space operation are exempt from the station identification requirements of §97.84 on each frequency band when in use.

(b) Stations in space operation may automatically retransmit the radio signals of other stations in earth operation, and space operation.

(c) Stations in space operation are exempt from the control operator requirements of §97.79 and from the provisions of §97.88 pertaining to the operation of a station by remote control

**§ 97.419 Telemetry.**

(a) Telemetry transmission by stations in space

[1]Stations operating in the Amateur-Satellite Service shall not cause harmful interference to other stations between 435 and 438 MHz. (See International Radio Regulations, RR MOD 3644/320A.)

operation may consist of specially coded messages intended to facilitate communications.

(b) Telemetry transmissions by stations in space operation are permissible one-way communications.

**§ 97.421 Telecommand operation.**

(a) Stations in telecommand operation may transmit special codes intended to obscure the meaning of command messages to the station in space operation.

(b) Stations in telecommand operation are exempt from the station identification requirements of §97.84.

(c) Stations in telecommand operation may transmit from within the military areas designated in §97.61(b)(7) in the frequency band 435-438 MHz with a maximum of 611 watts effective radiated power (1000 watts equivalent isotropically radiated power). The transmitting-antenna elevation angle between the lower half-power (– 3 decibels relative to the peak or antenna bore sight) point and horizon must always be greater than 10°.

**§ 97.422 Earth operation.**

Stations in earth operation may transmit from within the military areas designated in §97.61(b)(7) in the frequency band 435-438 MHz with a maximum of 611 watts effective radiated geographic area on the Earth's surface which is capable of being served by the station in space operation. Specify for both the transmitting and receiving antennas of this station.

*Orbital parameters.* A description of the anticipated orbital parameters as follows:

*Non-geostationary satellite*

1) Angle of inclination
2) Period
3) Apogee (kilometers)
4) Perigee (kilometers)
5) Number of satellites having the same orbital characteristics

*Geostationary satellite*

1) Nominal geographical longitude
2) Longitudinal tolerance
3) Inclination tolerance
4) Geographical longitudes marking the extremities of the orbital arc over which the satellite is visible at a minimum angle of elevation of 10° at points within the associated service area.
5) Geographical longitudes marking the extremities of the orbital arc within which the satellite must be located to provide communications to the specified service area
6) Reason when the orbital arc of (5) is less than that of (4)

*Technical Parameters.* A description of the proposed technical parameters for:

(1) the station in space operation; and

power (1000 watts equivalent isotropically radiated power). The transmitting-antenna elevation angle between the lower half-power (−3 decibels relative to the peak or antenna bore sight) point and the horizon must always be greater than 10°.

### § 97.423 Notification required.

(a) The licensee of every station in space operation shall give written notifications to the Private Radio Bureau, Federal Communications Commission, Washington, DC 20554.

(b) *Pre-space operation notification.*

(1) Three notifications are required prior to initiating space operation. They are:

*First notification.* Required no less than twenty-seven months prior to initiating space operation.

*Second notification.* Required no less than fifteen months prior to initiating space operation.

*Third notification.* Required no less than three months prior to initiating space operation.

(2) The pre-space operation notification shall consist of:

*Space operation date.* A statement of the expected date space operations will be initiated, and a prediction of the duration of the operation.

*Identity of satellite.* The name by which the satellite will be known.

*Service area.* A description of the

(2) a station in earth operation suitable for use with the station in space operation; and

(3) a station in telecommand operation suitable for use with the station in space operation.

The description shall include:

(1) Carrier frequencies if known; otherwise give frequency range where carrier frequencies will be located.

(2) Necessary bandwidth.

(3) Class of emission.

(4) Total peak power.

(5) Maximum power density (watts/Hz)

(6) Antenna radiation pattern[1]

[1]These antenna characteristics shall be provided for both transmitting and receiving antennas.

[2]For a station in space operation.

[3]The total noise temperature at the input of a typical amateur radio station receiver shall include the antenna noise (generated by external sources (ground, sky, etc.) peripheral to the receiving antenna and noise re-radiated by the satellite), plus noise generated internally to the receiver. The additional receiver noise is above thermal noise, $kT_oB$. Referred to the antenna input terminals, the total system noise temperature is given by

$$T_s = T_a + (L - 1) T_o + L T_r$$

where:

$T_a$: antenna noise temperature

L: line losses between antenna output terminals and receiver input terminals

$T_o$: ambient temperature, usually given as 290° K

$T_r$: receiver noise temperature, this is also given as $(NF-1)T_o$, where NF is receiver noise figure.

(7) Antenna gain (main beam)[1]

(8) Antenna pointing accuracy (geostationary satellites only)[1]

(9) Receiving system noise temperature[2]

(10) Lowest equivalent satellite link noise temperature[3]

(c) *In-space operation notification.* Notification is required after space operation has been initiated. The notification shall update the information contained in the pre-space notification. In-space operation notification is required no later than seven days following initiation of space operation.

(d) *Post-space operation notification.* Notification of termination of space operation is required no later than three months after termination is complete. If the termination is ordered by the Commission, notification is required no later than twenty-four hours after termination is complete.

# Subpart I — Volunteer-Examiner Coordinators

**General**

**§ 97.501 Purpose.**

The rules in this subpart are designed to provide for the establishment of volunteer-examiner coordinators to coordinate the efforts of volunteer examiners in preparing and administering examinations for amateur radio operator licenses.

**§ 97.503 Definitions.**

For the purpose of this subpart, the following definitions are applicable:

(a) *Volunteer-examiner coordinator* (VEC).

(6) California;

(7) Arizona, Idaho, Montana, Nevada, Oregon, Utah, Washington, and Wyoming;

(8) Michigan, Ohio and West Virginia;

(9) Illinois, Indiana and Wisconsin;

(10) Colorado, Iowa, Kansas, Minnesota, Missouri, Nebraska, North Dakota and South Dakota;

(11) Alaska;

(12) Carribean Insular areas: Commonwealth of Puerto Rico, United States Virgin Islands (50 islets and cays) and Navassa Island; *AND*

(13) Pacific Insular areas: Hawaii, American Somoa (seven islands), Baker Island, Com-

An organization which has entered into an agreement with the Federal Communications Commission to coordinate the efforts of volunteer examiners in preparing and administering examinations for amateur radio operator licenses.

(b) *Volunteer examiner*. An amateur radio operator who prepares or administers examinations to applicants for amateur radio operator licenses.

**§ 97.505 Applicability of rules.**

These rules apply to each organization that serves as a volunteer-examiner coordinator.

**§ 97.507 VEC Qualifications.**

In order to be a VEC, an organization must:

(a) Be organized at least partially for the purpose of furthering amateur radio;

(b) Be at least regional in scope, serving one or more of the following regions:

(1) Connecticut, Maine, Massachusetts, New Hampshire, Rhode Island and Vermont;

(2) New Jersey and New York;

(3) Delaware, the District of Columbia, Maryland and Pennsylvania;

(4) Alabama, Florida, Georgia, Kentucky, North Carolina, South Carolina, Tennessee and Virginia;

(5) Arkansas, Louisiana, Mississippi, New Mexico, Oklahoma and Texas;

monwealth of Northern Mariannas Islands, Guam Island, Howland Island, Jarvis Island, Johnston Island (Islets East, Johnston, North and Sand), Kingman Reef, Midway Island (Islets Eastern and Sand), Palmyra Island (more than 50 islets) and Wake Island (Islets Peale, Wake and Wilkes).

(c) Be capable of acting as a VEC in one or more of the regions listed in paragraph (b);

(d) Agree to coordinate all amateur radio operator examination elements for all amateur radio operator license classes except Novice Class;

(e) Agree not to accept any compensation from any source for its services as a VEC, except reimbursement for out-of-pocket expenses permitted by §97.36; and

(f) Agree to assure that for any examination every candidate qualified under these rules is registered without regard to race, sex, religion, national origin or membership (or lack thereof) in any amateur radio organization.

**§ 97.509 Conflicts of interest.**

An organization engaged in the manufacture or distribution of equipment used in connection with amateur radio transmissions, or in the preparation or distribution of any publication used in preparation for obtaining amateur radio station operator licenses may be a VEC only upon a persuasive showing to the Commission that preventive measures have been taken to preclude any possible conflict of interest.

## VOLUNTEER-EXAMINER COORDINATOR FUNCTIONS

### § 97.511 Agreement required.

No organization may serve as a VEC until that organization has entered into a written agreement with the Federal Communications Commission to do so. The VEC must abide by the terms of the agreement.

### § 97.513 Scheduling of examinations.

A VEC will coordinate the dates and times for scheduling examinations (see §97.26) throughout the region(s) it serves. Any VEC may also coordinate the scheduling of testing opportunities outside of the regions listed in Section 97.507(b). A VEC shall notify the Engineer-in-Charge of the Field Operations Bureau (FOB) District Office having jurisdiction over the area where an examination is to be held of the time, place and

(a) the volunteer examiner does not meet minimum statutory qualifications or minimum qualifications as prescribed by the rules;

(b) the FCC refuses to accept the voluntary and uncompensated services of the volunteer examiner;

(c) the VEC determines that the volunteer is not competent to perform the function for which he/she volunteered; *OR*

(d) the VEC determines that questions of the volunteer's integrity or honesty could compromise the examination(s).

### § 97.517 Examinations.

A VEC will design (see §97.27[d], assemble, print and distribute written examination Elements 3, 4(A) and 4(B). A VEC may design, assemble, print and distribute examination Elements 1(B) and 1(C). A VEC is required to hold examination designs in confidence.

### § 97.519 Examination procedures.

At the completion of each examination, a VEC will collect the candidates' application forms,

registration requirements for any examination. If no FOB District Office has jurisdiction over the area where an examination is to be held, a VEC shall notify the Chief of the Public Service Division of FOB in Washington, DC, instead. In either case, this notification must be made at least 30 days in advance of the registration deadline.

**§ 97.515 Coordinating Volunteer examiners.**

A VEC will accredit amateur radio operators, licensed by the Federal Communications Commission, as volunteer examiners (see §97.30). A VEC will seek to recruit a broad representation of amateur radio operators to be volunteer examiners. A VEC may not discriminate in accrediting volunteer examiners on the basis of race, sex, religion or national origin. A VEC may not refuse to accredit a volunteer on the basis of membership (or lack thereof) in an amateur radio organization. A VEC may not discriminate in accrediting volunteer examiners based upon their accepting or declining to accept reimbursement. A VEC must not accredit an amateur radio operator volunteering to be an examiner if:

answer sheets and test results from the volunteer examiners (see §97.28(h)). A VEC will:

(a) Make a record of the date and place of the test; the names of the volunteer examiners and their qualifications; the names of the candidates; the test results; and, related information.

(b) Screen the application for completeness and authenticity.

(c) Forward the application within ten days of its receipt from the examiners to: Federal Communications Commission, Licensing Division, Private Radio Bureau, Gettysburg, PA 17325.

(d) Make available to any authorized FCC representative any requested examination records.

**§ 97.521 Evaluation of questions.**

A VEC will be expected to evaluate the clarity and accuracy of examination questions on the basis of experience, and to bring ambiguous or inaccurate questions to the attention of the Commission, with a recommendation on whether to revise the question or to delete the question from the Commission's list of examination questions.

**§ 97.523 Reserved.**

# APPENDICES

## APPENDIX 1

### Examination Points

Examinations for Amateur Radio operator licenses are conducted at the Commission's office in Washington, DC, and at each field office of the Commission on the days designated by the Engineer-in-Charge of each office. Specific dates should be obtained from the Engineer-in-Charge of the nearest field office of the Commission.

Examinations are also given at prescribed intervals in the cities listed in the Commission's current Examination Schedule, copies of which are available from the Federal Communciations Commission Regional Services Division, Washington, DC 20554, or from any one of the Commission's field offices listed in §0.121.

## APPENDIX 2

*Extracts from the International Telecommunications Convention (Malaga-Torremolinois, 1973), as revised by the World Administrative Radio Conference, Geneva, 1979.*

**Article 1 — Terms and Definitions**

Section III. Radio Services

§3.34 *Amateur Service:*

shall be made in plain language and shall be limited to messages of a technical nature relating to tests and to remarks of a personal character for which, by reason of their unimportance, recourse to the public telecommunications service is not justified.

(2) It is absolutely forbidden for amateur stations to be used for transmitting international communications on behalf of third parties.

(3) The preceding provisions may be modified by special arrangements between the administrations of the countries concerned.

§3. (1) Any person seeking a license to operate the apparatus of an amateur station shall prove that he is able to send correctly by hand and to receive correctly by ear texts in Morse code signals. The administrations concerned may, however, waive this requirement in the case of stations making use exclusively of frequencies above 30 MHz.

(2) Administrations shall take such measures as they judge necessary to verify the operational and technical qualifications of any person wishing to operate the apparatus of an amateur station.

§4. The maximum power of amateur stations shall be fixed by the administrations concerned, having regard to the technical qualifications of the operators and to the conditions under which these stations are to operate.

A *radiocommunication service* for the purpose of self-training, intercommunication and technical investigations carried out by amateurs, that is, by duly authorized persons interested in radio technique solely with a personal aim and without pecuniary interest.

§3.35 *Amateur-Satellite Service:*

A *radiocommunicatons service* using *space stations* on earth *satellites* for the same purposes as those of the *amateur service.*

**Article 32 — Amateur Service and Amateur-Satellite Service**

Section I. Amateur Service

§1. Radiocommunications between amateur stations of different countries shall be forbidden if the administration of one of the countries concerned has notified that it objects to such radiocommunications.

§2. (1) When transmissions between amateur stations of different countries are permitted, they

§5. (1) All the general rules of the Convention and of these Regulations shall apply to amateur stations. In particular, the emitted frequency shall be as stable and as free from spurious emissions as the state of technical development for such stations permits.

(2) During the course of their transmissions, amateur stations shall transmit their call sign at short intervals.

Section II. Amateur-Satellite Service

§6. The provisions of Section I of this Article shall apply equally, as appropriate, to the Amateur-Satellite Service.

§7. Space stations in the Amateur-Satellite Service operating in bands shared with other services shall be fitted with appropriate devices for controlling emissions in the event that harmful interference is reported in accordance with the procedure laid down in Article 22. Administrations authorizing such space stations shall inform the IFRB and shall ensure that sufficient earth command stations are established before launch to guarantee that any harmful interference which might be reported can be terminated by the authorizing administration.

[1]As may appear in public notices issued by the Commission.

## APPENDIX 3

### Classification of Emissions

For convenient reference the tabulation below is extracted from the classification of typical emissions in Part 2 of the Commission's Rules and Regulations. It includes only those general classifications which appear most applicable to the Amateur Radio Service.

| Type of modulation | Type of transmission | Symbol |
|---|---|---|
| Amplitude | With no modulation | AØ |
| | Telegraphy without the use of modulating audio frequency (by on-off keying) | A1 |
| | Telegraphy by the on-off keying of an amplitude modulating audio frequency or audio frequencies or by the on-off keying of the modulated emission (special case: | A2 |
| | Telephone | F3 |
| | Facsimile | F4 |
| | Television | F5 |
| Pulse | | P |

## APPENDIX 4

*Convention Between Canada and the United States of America, Relating to the operation by Citizens of Either Country of Certain Radio Equipment or Stations in the Other Country (Effective May 15, 1952)*

### Article III

It is agree that persons holding appropriate amateur licenses issued by either country may operate their amateur stations in the territory of the other country under the following conditions:

(a) Each visiting amateur may be required to register and receive a permit before operating any amateur station licensed by his government.

(b) The visiting amateur will identify his station by:

(1) *Radiotelegraph operation.* The amateur call sign issued to him/her by the licensing country followed by a slant (/) sign and the amateur call sign prefix and call area number of the country he is visiting.

(2) *Radiotelephone operation.* The amateur

| | | |
|---|---|---|
| | an unkeyed emission amplitude modulated). | |
| | Telephony | A3[1] |
| | Facsimile | A4 |
| | Television | A5 |
| Frequency (or phase) | Telegraphy by frequency shift keying without the use of a modulating audio frequency. | F1 |
| | Telegraphy by the on-off keying of a frequency modulating audio frequency or by the on-off keying of frequency modulated emission (special case; an unkeyed emission frequency modulated). | F2 |

[1](In Part 97) Unless specified otherwise, A3 includes single and double sideband with full, reduced or suppressed carrier.

call sign in English issued to him by the licensing country followed by the words, "fixed" "portable" or "mobile," as appropriate, and the amateur call sign prefix and call area number of the country he is visiting.

(c) Each amateur station shall indicate at least once during each contact with another station its geographical location as nearly as possible by city and state or city and province.

(d) In other respects the amateur station shall be operated in accordance with the laws and regulations of the country in which the station is temporarily located.

## APPENDIX 5

The effective height of the transmitting antenna shall be the height of the antenna's center of radiation above "average terrain." For this purpose "effective height" shall be established as follows:

(a) On a U.S. Geological Survey Map having a scale of 1:250,000, lay out eight evenly spaced radials, extending from the transmitter site to a distance of 10 miles and beginning at (0°, 45°, 90°, 135°, 180°, 225°, 270°, 315° T.) If preferred, maps of greater scale may be used.

(b) By reference to the map contour lines, establish the ground elevation above mean sea level (AMSL) at 2, 4, 6, 8, and 10 miles from the antenna structure along each radial. If no elevation figure or contour line exists for any particular point, the nearest contour line elevation shall be employed.

(c) Calculate the arithmetic mean of these 40 points of elevation (5 points of each of 8 radials).

(d) The height above average terrain of the antenna is thus the height AMSL of the Antenna's center of radiation, minus the height of average terrain as calculated above.

NOTE 1: Where the transmitter is located near a large body of water, certain points of established elevation may fall over water. Where it is expected that service would be provided to land areas beyond the body of water, the points at water level in that direction should be included in the calculation of average elevation. Where it is expected that service would not be provided to land areas beyond the body of water, the points at water level should not be included in the average.

NOTE 2: In instances in which this procedure might provide unreasonable figures due to the unusual nature of the local terrain, applicant may provide additional data at his own discretion, and such data may be considered if deemed significant.

widespread distribution and their demonstrated capacity in such cases, can assist in meeting essential communication needs.

f) that existence of national and regional amateur emergency networks using frequencies throughout the bands allocated to the amateur service.

g) that in the event of a natural disaster, direct communications between amateur stations and other stations might enable vital communications to be carried out until normal communications are restored.

*Recognizing*

that the rights and responsibilities for communications in the event of a natural disaster rest with the administrations involved.

*Resolves*

1. that the bands allocated to the amateur service which are specified in No. 510 may be used by administrations to meet the needs of international disaster communications.
2. that such use of these bands shall be only for communications in relation to relief operations in connection with natural disasters.
3. that the use of specified bands allocated to the amateur service by non-amateur stations for

## APPENDIX 6

*Extracts from the International Telecommunicatons Convention (Malaga-Torremolinos, 1973), as revised by the World Administrative Radio Conference, Geneva, 1979.*

Resolution No. 640

Relating to the International Use of Radiocommunications, in the Event of Natural Disasters, in Frequency Bands Allocated to the Amateur Service.

*Considering*

a) that in the event of natural disaster normal communication systems are frequently overloaded, damaged, or completely disrupted.

b) that rapid establishment of communications is essential to facilitate worldwide relief actions.

c) that the amateur bands are not bound by international plans or notification procedures, and are therefore well adapted for short-term use in emergency cases.

d) that international disaster communications would be facilitated by temporary use of certain frequency bands allocated to the amateur service.

e) that under those circumstances the stations of the amateur service because of their disaster communications shall be limited to the duration of the emergency and to the specific geographical areas as defined by the responsible authority of the affected country.

4. that disaster communications shall take place within the disaster area and between the disaster area and the permanent headquarters of the organization providing relief.

5. that such communications shall be carried out only with the consent of the administration of the country in which the disaster has occurred.

6. that relief communications provided from outside the country in which the disaster has occurred shall not replace existing national or international amateur emergency networks.

7. that close cooperation is desirable between amateur stations and the stations of other radio services which may find it necessary to use amateur frequencies in disaster communications.

8. that such international relief communications shall avoid, as far as practicable, interference to the amateur service networks.

*Invites Administrations*

1. to proivde for the needs of international disaster communications.

2. to provide for the needs of emergency communications within their national regulations.

## APPENDIX 7

*Extracts from the International Telecommunications Convention (Malaga-Torremolinos, 1973), as revised by the World Administrative Radio Conference, Geneva, 1979.*

Resolution No. 641

Relating to the use of the Frequency Band 7000-7100 kHz

*Considering*

a) that the sharing of frequency bands by amateur and broadcasting services is undesirable and should be avoided.

b) that it is desirable to have worldwide exclusive allocations for these services in Band 7.

c) that the band 7000-7100 kHz is allocated on a worldwide basis exclusively to the amateur service.

*Resolves*

the Amateur-Satellite Service vary widely.

b) that space stations in the Amateur-Satellite Service are intended for multiple access by amateur earth stations in all countries.

c) that coordination among stations in the amateur and Amateur Satellite Services is accomplished without the need for formal procedures.

d) that the burden of terminating any harmful interference is placed upon the administration authorizing a space station in the Amateur-Satellite Service pursuant to the provisions of No. 2741 of the Radio Regulations.

*Notes*

that certain information specified in Appendices 3 and 4 cannot reasonably be provided for earth stations in the Amateur-Satellite Service.

*Resolves*

1. that when an administration (or one acting

that the broadcasting service shall be prohibited from the band 7000-7100 kHz and that the broadcasting stations operating on frequencies in this band shall cease such operation.

## APPENDIX 8

*Extracts from the International Telecommunications Convention (Malaga-Torremolinos, 1973), as revised by the World Administrative Radio Conference, Geneva, 1979.*

Resolution No. 642

Relating to the Bringing into Use of Earth Stations in the Amateur-Satellite Service.

*Recognizing*

that the procedures of Articles 11 and 13 are applicable to the Amateur-Satellite Service.

*Recognizing Further*

a) that the characteristics of each station in on behalf of a group of named administrations) intends to establish a satellite system in the Amateur-Satellite Service and wishes to publish information with respect to earth stations in the system it may:

1.1 communicate to the IFRB all or part of the information listed in Appendix 3; the IFRB shall publish such information in a special section of its weekly circular requesting comments to be communicated within a period of four months after the date of publication.

1.2 notify under Nos. 1488 to 1491 all or part of the information listed in Appendix 3; the IFRB shall record it in a special list.

2. that this information shall include at least the characteristics of a typical amateur earth station in the Amateur-Satellite Service having the facility to transmit signals of the space station to initiate, modify, or terminate the functions of the space station.

# Appendix C
# FCC Field Offices

Listed below are the addresses and telephone numbers of the FCC field offices. This list is alphabetical by state, and also includes the field offices in Puerto Rico and The District of Columbia, (Washington, D.C.).

**Alaska, Anchorage**

U.S. Post Office Building, Room G63
4th & G Street, P.O. Box 644
Anchorage, Alaska 99510
Phone: Area Code 907-265-5201

**California, Long Beach**

3711 Long Beach Blvd.
Suite 501
Long Beach, California 90807
Office Examinations (Recording)
Phone: Area Code 213-426-7886
Other Information
Phone: Area Code 213-426-4451

**California, San Diego**

Fox Theatre Building
1245 Seventh Avenue
San Diego, California 92101
Office Examinations (Recording)
Phone: Area code 714-293-5460
Other Information
Phone: Area Code 714-293-5478

**California, San Francisco**

323A Customhouse
555 Battery Street
San Francisco, California 94111
Office Examinations (Recording)
Phone: Area Code 415-556-7700
Other Information
Phone: Area Code 415-556-7701

**Colorado, Denver**

Suite 2925, The Executive Tower
1405 Curtis Street
Denver, Colorado 80202
Office Examinations (Recording)
Phone: Area Code 303-837-4053
Other Information
Phone: Area Code 303-837-5137

**District of Columbia (Washington, D.C.)**

1919 M Street N.W. Room 411
Washington, D.C. 20554
Phone: Area Code 202-632-8834

**Florida, Miami**

919 Federal Building
51 S.W. First Avenue
Miami, Florida 33130
Phone: Area Code 305-350-5541

**Florida, Tampa**

809 Barnett Bank Building
1000 Ashley Street
Tampa, Florida 33602
Office Examinations (Recording)
Phone: Area Code 813-228-2605
Other Information
Phone: Area code 404-881-3084

**Georgia, Atlanta**

Room 440, Massell Building
1365 Peachtree St. N.E.
Atlanta, Georgia 30309
Office Examinations (Recording)
Phone: Area Code 404-881-7381
Other Information
Phone: Area Code 404-881-3084

**Georgia, Savannah**

238 Federal Office Building
and Courthouse
125 Bull Street, P.O. Box 8004
Savannah, Georgia 31402
Phone: Area Code 912-232-4321
ext. 320

**Hawaii, Honolulu**

7304 Prince Jonah Kuhio
Kalanianaole Building
300 Ala Moana Blvd.
Honolulu, Hawaii 96813
Phone: Area Code 808-546-5640

**Illinois, Chicago**

3935 Federal Building
230 South Dearborn Street
Chicago, Illinois 60604
Office Examinations (Recording)
Phone: Area Code 312-353-0197
Other Information
Phone: Area Code 312-353-0195

**Louisiana, New Orleans**

829 F. Edward Hebert Federal Building
600 South Street
New Orleans, Louisiana 70130
Phone: Area code 504-589-2094

**Maryland, Baltimore**

George M. Fallon Federal Building
Room 819 31 Hopkins Plaza
Baltimore, Maryland 21201
Office Examinations (Recording)
Phone: Area Code 301-962-2727
Other Information
Phone: Area Code 301-962-2728

**Massachusetts, Boston**

1600 Customhouse
165 State Street
Boston, Massachusetts 02109
Office Examinations (Recording)
Phone: Area Code 617-223-6608
Other Information
Phone: Area Code 617-223-6609

**Michigan, Detroit**

1054 Federal Building &
U.S. Courthouse
231 W. Lafayette Street
Detroit, Michigan 48226
Office Examinations (Recording)
Phone: Area Code 313-226-6077
Other Information
Phone: Area Code 313-226-6078

**Minnesota, St. Paul**

691 Federal Building
316 N. Robert Street
St. Paul, Minnesota 55101
Office Examinations (Recording)
Phone: Area Code 612-725-7819
Other Information
Phone: Area Code 612-725-7810

**Missouri, Kansas City**

1703 Federal Building
601 East 12th Street
Kansas City, Missouri 64106
Office Examinations (Recording)
Phone: Area Code 816-374-5526
Other Information
Phone: Area Code 816-374-6155

**New York, Buffalo**

1307 Federal Building
111 W. Huron Street at Delaware Ave.
Buffalo, New York 14202
Phone: Area Code 716-842-3216

**New York, New York**

201 Varick Street
New York, New York 10014
Office Examinations (Recording)
Phone: Area Code 212-620-3435
Other Information
Phone: Area Code 212-620-3437

**Oregon, Portland**

1782 Federal Office Building
1220 S.W. 3rd Ave.
Portland, Oregon 97204
Office Examinations (Recording)
Phone:
Area Code 503-221-3097
Other Information
Phone:
Area Code 503-221-3098

**Pennsylvania, Philadelphia**

11425 James A. Byme Federal Courthouse
601 Market Street
Philadelphia, Pennsylvania 19106
Office Examinations (Recording)
Phone: Area Code 215-597-4410
Other Information
Phone: Area Code 215-597-4411

**Puerto Rico, Hato Rey (San Juan)**

Federal Building & Courthouse, Room 747
Avenida Carlos Chardon
Hato Rey, Puerto Rico 00918
Phone: Area Code 809-753-4567 or
Phone: Area Code 809-753-4008

**Texas, Beaumont**

Room 323 Federal Building
300 Willow Street
Beaumont, Texas 77701
Phone: Area Code 713-838-0271, Ext. 317

**Texas, Dallas**

Earle Cabell Federal Bldg.
Room 13E7, 1100 Commerce St.
Dallas, Texas 75242
Office Examinations (Recording)
Phone:
Area Code 214-749-3243
Other Information
Phone:
Area Code 214-749-1719

**Texas, Houston**

5636 Federal Building
515 Rusk Avenue
Houston, Texas 77002
Office Examinations (Recording)
Phone:
Area Code 713-226-4306
Other Information
Phone:
Area Code 713-226-5624

**Virginia, Norfolk**

Military Circle
870 North Military Highway
Norfolk, Virginia 23502
Office Examinations (Recording)
Phone: Area Code 804-461-4000
Other Information
Phone: Area Code 804-441-6472

**Washington, Seattle**

3256 Federal Building
915 Second Ave.
Seattle, Washington 98174
Office Examinations (Recording)
Phone: Area Code 206-442-7610
Office Information
Phone: Area Code 206-442-7653

# Index

# Index

## W

# Other Bestsellers From TAB

☐ **INTRODUCTION TO TELECOMMUNICATIONS SYSTEMS—P.H. Smale**

This sourcebook covers the whole range of telecommunication principles, beginning with a basic discussion of wave theory and continuing with the principles behind radio, television, telephone, and digital networks. Covers all the latest technology in telecommunications: local area networks (LANs), cellular phones, cable television, direct broadcasting by satellite, high-definition television, optic fiber systems, and mobile radio systems. 290 pp., 181 illus.

**Paper $14.95 Book No. 2924**

☐ **THE PACKET RADIO HANDBOOK—Jonathan L. Mayo, KR3T**

Gain access to a whole new communications experience by using a microcomputer to digitally encode information that is then transmitted via radio to another packet radio station. If you've heard about this fast-paced, state-of-the-art operation, or if you already own packet radio equipment and would like to learn how to better utilize it . . . this one-of-a-kind guide will show you how to get in on the action! Also covered is modulation techniques, protocols, hardware development and networks. 224 pp., 91 illus.

**Paper $14.95 Book No. 2722**

☐ **THE COMPLETE SHORT-WAVE LISTENER'S HANDBOOK—3rd Edition—Hank Bennett, Harry L. Helms, and David T. Hardy**

Now this classic sourcebook has been revised and updated to include all the latest innovations in SWL equipment and operating practices. Covers everything from terminology, signals, antennas, broadcast bands, harmonics, and reception areas, to VHF and UHF monitoring, equipment terms, tips on keeping a logbook, plus advice on listening to pirate stations, RTTY, Air Force One, and more! 304 pp., 96 illus.

**Paper $16.95 Book No. 2655**

☐ **CELLULAR TELEPHONES—A LAYMAN'S GUIDE—Stuart Crump, Jr.**

With this handbook as your guide, you can become part of the cellular phone revolution at a cost you can afford! It tells you everything you need to know including how to use your cellular phone as a personal and business productivity tool . . . how to make the most efficient use of your mobile phone . . . even how to get the maximum status value out of your new car phone. 160 pp., 42 illus.

**Hard $15.95 Book No. 1965**